ADVANCE PRAISE FOR *SAVING THE NEWS*

"Dramatic technological and economic change threaten the information infrastructures that democracy requires to survive. Martha Minow provides a legal and policy roadmap for a 21st century media ecosystem. As we face a global crisis of democracy, Minow's analysis and proposals are more timely and urgent than ever. A must-read."

—**K. Sabeel Rahman**, President of Demos, and Associate Professor of Law, Brooklyn Law School

"Addressing one of the most vexing challenges of our time, Martha Minow has written the essential book on the topic. Easy to say that something should be done about the affect of tech and social media on news and civic discourse. Much harder to say what to do. This brilliant, illuminating and much-needed book moves the discussion toward solutions. *Saving the News* is thoughtful and thought-provoking, and a must-read."

—**Julius Genachowski**, Managing Director of global technology, media and telecommunications, The Carlyle Group, and former Chairman of the US Federal Communications Commission

SAVING THE NEWS

OTHER BOOKS IN THE SERIES:

Not a Suicide Pact
The Constitution in a Time of National Emergency
Richard A. Posner

Out of Range
Why the Constitution Can't End the Battle over Guns
Mark V. Tushnet

Unfinished Business
Racial Equality in American History
Michael J. Klarman

Supreme Neglect
How to Revive Constitutional Protection for Private Property
Richard A. Epstein

Is There a Right to Remain Silent?
Coercive Interrogation and the Fifth Amendment After 9/11
Alan M. Dershowitz

The Invisible Constitution
Laurence H. Tribe

Uninhibited, Robust, and Wide-Open
A Free Press for a New Century
Lee C. Bollinger

From Disgust to Humanity
Sexual Orientation and Constitutional Law
Martha C. Nussbaum

The Living Constitution
David A. Strauss

Keeping Faith with the Constitution
Goodwin Liu, Pamela S. Karlan, and Christopher H. Schroeder

Cosmic Constitutional Theory
Why Americans Are Losing Their Inalienable Right to Self-Governance
J. Harvie Wilkinson III

More Essential than Ever
The Fourth Amendment in the Twenty-First Century
Stephen J. Schulhofer

On Constitutional Disobedience
Louis Michael Seidman

The Twilight of Human Rights Law
Eric A. Posner

Constitutional Personae
Heroes, Soldiers, Minimalists, and Mutes
Cass R. Sunstein

The Future of Foreign Intelligence
Privacy and Surveillance in a Digital Age
Laura K. Donohue

HATE
Why We Should Resist It with Free Speech, Not Censorship
Nadine Strossen

Democracy and Equality
The Enduring Constitutional Vision of the Warren Court
Geoffrey R. Stone and David A. Strauss

Beyond Belief, Beyond Conscience
The Radical Significance of the Free Exercise of Religion
Jack N. Rakove

The Religion Clauses
The Case for Separating Church and State
Erwin Chemerinsky and Howard Gilman

Liars
Falsehoods and Free Speech in an Age of Deception
Cass R. Sunstein

INALIENABLE RIGHTS SERIES

. . .

David A. Strauss
Gerald Ratner
Distinguished Service Professor of Law
University of Chicago Law School

Kathleen M. Sullivan
Stanley Morrison Professor of Law
Stanford Law School

Cass R. Sunstein
Robert Walmsley University Professor
Harvard Law School

Laurence H. Tribe
Carl M. Loeb University
Professor of Law
Harvard Law School

Mark V. Tushnet
William Nelson Cromwell
Professor of Law
Harvard Law School

J. Harvie Wilkinson III
Judge
U.S. Court of Appeals for the Fourth Circuit

Kenji Yoshino
Chief Justice Earl Warren
Professor of Constitutional Law
New York University School of Law

Geoffrey Stone and Oxford University Press gratefully acknowledge the interest and support of the following organizations in the Inalienable Rights series: The ALA The Chicago Humanities Festival The American Bar Association The National Constitution Center The National Archives

Saving the News

Why the Constitution Calls for Government Action to Preserve Freedom of Speech

Martha Minow

OXFORD
UNIVERSITY PRESS

OXFORD
UNIVERSITY PRESS

Oxford University Press is a department of the University of Oxford. It furthers the University's objective of excellence in research, scholarship, and education by publishing worldwide. Oxford is a registered trade mark of Oxford University Press in the UK and certain other countries.

Published in the United States of America by Oxford University Press
198 Madison Avenue, New York, NY 10016, United States of America.

Library of Congress Cataloging-in-Publication Data
Names: Minow, Martha, 1954– author.
Title: Saving the news : why the Constitution calls for government action to preserve freedom of speech / Martha Minow.
Description: New York : Oxford University Press, [2021] |
Series: Inalienable rights series | Includes index. |
Identifiers: LCCN 2020056753 (print) | LCCN 2020056754 (ebook) |
ISBN 9780190948412 (hardback) | ISBN 9780190948436 (epub) |
ISBN 9780190948443
Subjects: LCSH: Press law—Economic aspects—United States. |
Freedom of the press—United States.
Classification: LCC KF2750 .M56 2021 (print) | LCC KF2750 (ebook) |
DDC 342.7308/53—dc23
LC record available at https://lccn.loc.gov/2020056753
LC ebook record available at https://lccn.loc.gov/2020056754

DOI: 10.1093/oso/9780190948412.001.0001

1 3 5 7 9 8 6 4 2

Printed by LSC communications, United States of America

To M.S. and other intrepid witnesses and advocates for accountable democracy

If you want to preserve democracy as we know it, you have to have a free and many times adversarial press. And without it, I am afraid that we would lose so much of our individual liberties over time. That's how dictators get started.

—Senator John McCain, February 20, 2017

Contents

. . .

Preface

. . .

FROM GUTENBERG TO ZUCKERBERG

By Newton Minow

When Johannes Gutenberg invented movable type and the printing press in the fifteenth century, he married print and paper. Five centuries later, Mark Zuckerberg and others caused the long marriage of print and paper to stumble. Technology now moves faster than ever before. The internet eliminates distance and boundaries. Artificial intelligence looms to replace human judgment in the next step in the global communications revolution.

Our daughter Martha asked me to lend perspective to how changing technology is impacting communications policy. Over the last seven decades, I've been privileged to have a front-row seat inside the communications revolution. It began in World War II when I was a teenage U.S. Army sergeant in the 835th Signal Service Battalion in the China-Burma-India Theater. Our battalion built the first telephone line along the Burma (now Myanmar) Road connecting India with China. After college and law school, I was a law clerk for

Chief Justice Fred Vinson when the Supreme Court of the United States heard a case that fascinated me: it was about how the Federal Communications Commission (FCC) reached its decision regarding competing technological standards for color television. After my tenure with the chief justice, I became assistant counsel for Illinois governor Adlai E. Stevenson and worked in his two campaigns for president in 1952 and 1956. One of my assignments was to appeal to the broadcast networks and to the FCC for equal time for Stevenson on radio and television so that Stevenson could respond to President Eisenhower. In 1961, President Kennedy appointed me chairman of the FCC. Returning in 1963 to private life and law practice, I became a managing partner of what is now an international law firm (Sidley Austin) in the United States, Europe, and Asia, its offices tied together by the most advanced digital technology. In the next fifty years, I served on many nonprofit and for-profit boards of directors, including for a major book publisher (Encyclopaedia Britannica), a major magazine publisher (Curtis Publishing, which produced the *Saturday Evening Post*), major newspapers (*Chicago Sun-Times*, *Chicago Daily News*, *Chicago Tribune*), independent UHF television stations (Field Enterprises), public television (WTTW Chicago, PBS), a national radio and television network (CBS), an international advertising agency (Foote, Cone & Belding/Publicis), a major advertiser (Sara Lee), a major think tank (RAND Corporation, which had a huge role in creating the internet), a major philanthropic foundation (the Carnegie Corporation of New York, which funded *Sesame Street*), the televised presidential debates (first with the League of Women Voters, then with the Commission on Presidential Debates), a communications policy leader (the Annenberg Washington Program), the world's most advanced telemedicine service (the Mayo Clinic), and two major universities (Northwestern and Notre Dame); I also taught graduate journalism and law students (at Northwestern University). These experiences gave me diverse perspectives. I also served as chair of a special bipartisan advisory committee to the secretary of defense on protecting civil liberties in the fight against terrorism. I've seen every side of the elephant, including the backside.

Because of these experiences, Martha asked me to write this preface for her latest book. Martha grew up with nightly dinner conversations around the table with her mother, sisters, and me, and she listened to many stories and asked many questions. Martha asked me to describe what I learned. Now in my ninety-fifth year, I have five reflections.

First, two words—public interest—are disappearing from communications policy. When our government began to regulate communications about a hundred years ago, these two words set the standards. They appear in the Federal Communications Act dozens of times. Other nations, especially the United Kingdom, Canada, Japan, and Australia, use similar words to establish standards that broadcasters, cable operators, telephone companies, and other communicators must serve. Original regulatory systems were based on the idea that telephone service would be provided by wire and broadcast signals would travel through public airwaves. Today, most telephone calls originate and arrive through the air and most television viewers watch programs through wires. As technology changes, public policy lags behind. And the basic concept that our communications systems are to serve the public—not private—interest is now missing in action.

Second, changes in communications technology change not only the lives of individuals but also the roles of institutions, including governmental institutions. When I went to the FCC, I saw that the television and advertising industries paid no attention to official United States maps of cities, states, and legal boundaries. They created their own maps because radio and television signals do not respect traditional boundaries. Signals travel in a roughly sixty-mile circle, and viewers within that circle live in different cities, states, and rural areas. So instead of a map of forty-eight states in the continental United States, there is a map of 210 DMAs (Designated Market Areas). For example, a television transmitter in Chicago sends signals not only into Illinois but also to parts of Indiana, Wisconsin, Michigan, and Iowa. Without our knowing it, technology thus amends laws and constitutions. Our governmental institutions lag far behind changing technology.

Third, when I was at the FCC, we believed that the American people would benefit from more choice in radio and television service. We opened up FM radio, UHF television, cable television, public television, satellite television, and subscription television—greatly enlarging the existing 1961 television service of two and a half commercial networks. We added hundreds of new local stations. On reflection today, I wonder if enlarging choice contributed heavily to the deep divisions in our country, which is now more divided than I've ever seen before. In 1963 and in 2001, radio and television united our country in times of crisis, such as after the assassination of President Kennedy and the 9/11 terrorist attack. Americans today are divided not only on what they believe but also on what they "know," presenting not just different ideas but different facts. Walter Cronkite and I served on the CBS board together when he was the most trusted man in America. Now, who is trusted? How do we restore faith in facts? As Pat Moynihan said, we are all entitled to our own opinions, but not to our own facts. Although I still believe we were right to enlarge choice, I am no longer so sure.

Fourth, looking to the future, artificial intelligence (AI) is already well developed, and AI and the Internet of Things will soon revolutionize what we do and how we live. The lightning speed of AI can transform human reasoning and decision-making. Already we have observed AI defeating the world's master chess players and

Chinese experts at Go, the world's hardest game of psychological strategy. AI is determining which citizens get parole. Experts predict that AI will help guide its own evolution. We have been so busy learning how to use technology that we neglected to learn how to direct and govern technology. New and scary advances in AI enable what are called "deepfakes." AI can change a video by substituting any face wanted for the face of the speaker, can change the words of the speaker, and even change the way the speaker's lips originally moved. Political campaigns and elections can be manipulated by such deepfakes to threaten the future of democracy. No scientist, philosopher, or engineer has yet figured out how to program AI to serve the public interest.

Fifth, the result of all these changes is a profound challenge to democracy. As many scholars have written, our communications law and policy have long been based on the notion of a world in which speech is scarce and audiences abundant; today speech is abundant and listener attention is scarce. Unwittingly, we have so democratized the speech market that no one can be heard, bad actors flood social media, and democratic deliberation is damaged. That market has also created a state of constant information surveillance that threatens basic values of free expression.

Just as representative democracy is threatened by these changes, so is the international system that has sustained world peace and cooperation since the end of World War II. That system always had democratic deficiencies, but it worked. What will replace it? The populist movements sweeping liberal states have lots of sources, but all depend to some degree for their energy on the new social media and the capture of traditional media. Nowhere in this new order is there a thoughtful consideration of the public interest.

These changes should remind us of the words of Edward R. Murrow, who spoke about television in its earliest days: "This instrument can teach, it can illuminate; yes, and even it can inspire. But it can do so only to the extent that humans are determined to use it to those ends. Otherwise, it's nothing but wires and lights in a box." Martha is one of those determined humans—and that is why her book is so important.

Introduction

Jeopardy to News Production and Challenges for Constitutional Democracy

The United States Constitution specifically mentions only one private enterprise—the press—and does so in the context of according it constitutional protection.[1] A press free to criticize those in power and to spread information about developments, challenges, and opportunities across society figured high in the understanding of the Constitution's framers as they sought to create foundations for a strong democratic government and society. What does and what should the constitutional guarantee of press freedom mean at a time when the for-profit basis of newsgathering and sharing is strained, even failing? Does that guarantee, permit, prohibit, or require government steps to keep the press and its news work viable? There have been warnings for many decades now about the economic fragility of the business of gathering and sharing news. As outlets increased staff reductions and more newspapers closed across the United States in the first decades of the twenty-first century, the warnings escalated.[2]

Making sense of the trends is a complicated task. With the development of digital resources beginning in the late twentieth century, this age is awash with communications. Over the past centuries of practice, "the news" evolved to include reports of events, data, information, facts, and analyses offering informed and trustworthy communication of happenings, trends, and issues affecting people's lives. In the internet age, amid this plenitude are three features that depart from traditional techniques of newsgathering and presentation.

First, the sheer volume of material—including items from misinformation campaigns—makes it much more difficult for individuals to find and understand news that may matter to them. Second, current disruptions undermine the virtuous cycle in which news reporting grew with financial returns from subscriptions and sale of advertising, which in turn attracted more readers, subscriptions, and ads. Now, many readers have migrated to digital platforms that do not reinvest in reporting and analyzing news and do not see themselves as news providers. Leaders of digital companies have pretended that their platforms make no editorial choices for which they should be responsible. Third, and relatedly, disinvestment in newsgathering and reporting leaves often sizable gaps.

Those gaps are especially notable in the loss of local news operations in the United States, where lack of reporting about towns, suburbs, and rural areas is now creating "news deserts" across the country.[3] Changes in the private industry of the press leave some communities with no local news coverage.[4] In the local news outlets that remain, fewer than 20 percent of the stories deal with the community or events that take place there.[5] There has been less disruption of smaller newspapers by the national digital platforms than of bigger news operations, but consolidation of ownership and cost cutting have diminished coverage of local news. The new owners of newspapers and big digital platforms can choose not to invest in news production or what might be called "local government accountability" (reporting on the behavior of local officials or the state of local health, safety, economic, or education conditions). Traditional news media have shrunk, cutting staff and relying on freelancers, as digital platforms have surged. Fewer than one-third of people surveyed in 2016 trusted mass media to report news fully and accurately.[6] By 2020, 68 percent of Americans polled see too much bias in news reporting as a major problem—and 81 percent identified news media as critical or very important to democracy.[7] The decline in local news coverage may be tied to falling voter turnout for state and local elections and fewer people running for office in local elections.[8]

Newspaper newsrooms lost 45 percent of their employees between 2008 and 2017. Cascading reductions of staff and cutbacks on production accelerated in 2020 in the wake of the COVID-19 pandemic.[9] With revenue coming largely from advertising that targets individuals based on their identities and interests, digital media can easily sort people into different subcommunities, where they encounter different versions of events and concerns (especially so for those who are most politically engaged). Targeted marketing and algorithms divide people into subgroups in what might be called "digital gerrymandering," leading to quite different news, agendas, "facts," and understandings. Rather than coming across a variety of stories and viewpoints, individuals receive materials reflecting their past interests; predictions of interest based on their demographic, purchasing, and viewing habits; and nudges into content that will keep them on the site even if that content is inaccurate or extreme. As a result, large numbers of people live in worlds with barely overlapping news streams. The declining role of professional journalists and the vulnerability of digital platforms to invasion by foreign actors, bots, and manipulative interests contribute both to distrust of media and to misinformation, harming the efficacy of self-governance.

A majority of people in the United States now receive news selected for them by a computer-based mathematical formula derived from their past interests, producing echo chambers with few opportunities to learn, understand, or believe what others are hearing as news. People of course use digital platforms for many other purposes beyond getting news—entertainment, posting personal photographs, and so on. But with the shift of attention away from newspapers and broadcasting, advertising dollars too move to digital platforms.[10] The long-standing model of for-profit newspapers supported largely by advertising does not work when 89 percent of the online advertising dollars go to Google or Facebook and 60–70 percent of all advertising revenues go to internet companies. This trend will continue because digital ads are cheaper and aim with greater accuracy at likely customers.

These changes reflect a decline from the golden age of journalism, 1960–1980. When multiple news producers in cities treat news reporting as a public good, its value does not diminish as more people consume it, but people can "free ride" on its existence without paying for it, as it is costly or impossible to exclude them all. Knowledge is a classic public good; broadcast signals for radio and television resemble a public good, but cable providers found ways to exclude those who do not pay. Newspapers and broadcasters depend largely on commercial advertising, an economic path now profoundly diminished by the migration of ads to the digital sphere.[11]

The press has had earlier phases lacking elements of professional rigor. In the early nineteenth century, newspapers were organized and financed by political parties. In the late nineteenth century, "yellow journalism" pushed scandals. And U.S. history has repeatedly seen older forms of communication disrupted by new ones, as the telegraph and then broadcasting challenged newspapers. But the surmounting of past challenges does not erase the risk that journalism in the early twenty-first century will fall into even more dire straits.

The founders of the United States understood the central role played by the press in the American Revolution and as a guard against tyrannical government.[12] The Bill of Rights, amending the Constitution in response to many concerned about a central government being too powerful, embraced not only freedom of speech for individuals but a specific guarantee of freedom of the press. As crafted, the First Amendment of the U.S. Constitution assumes the existence and durability of a private industry. In directing that Congress "shall make no law . . . abridging the freedom of . . . the press," the Constitution's authors assumed the existence of newspapers.[13] Newspapers were, at that time, produced entirely on privately owned printing presses; they published accounts of events of the day, political opinions, essays, and entertainment for readers.[14] Constitutional protection for this work could have fallen comfortably within the legal protection of private property or individuals' freedom of speech. But the authors and voters behind the First Amendment thought the press important enough to single it out as a distinct bulwark for the liberty of the people and their vision of self-government. The shift from printing presses and delivery boys to tubes and fibers does not matter. Jeopardy to the very project of gathering and sharing actual news does.

The First Amendment does not govern the conduct of entirely private enterprises, but nothing in the Constitution forecloses government action to regulate concentrated economic power, to require disclosure of who finances communications, or to support news initiatives where market failures exist. Despite some current constitutional objections to any kind of governmental involvement in regulation of the press, there is a long-standing history of such involvement in the United States that undermines the purely constitutional aspects of these arguments.

The federal government has contributed financial resources, laws, and regulations to develop and shape media in the United

States. The transformation of media from printing presses to the internet was affected by deliberate government policies that influenced the direction of private enterprise. From granting newspapers low mailing rates (and even exemptions from postal fees) to investments in research (ultimately producing the internet), and from establishing licensing regimes for broadcasting to regulating telephone lines and features of digital platforms, the government has crafted the direction and contours of America's media ecosystem. The federal government has invested in the development of a new medium, shielded innovative media from competition, and used competition rules, subsidies, and other policies to promote access and innovation.

Indeed, the large degree of government involvement in media, in combination with media's functional importance to democracy, lends constitutional significance to policy choices present today. Potential reforms include a new fairness doctrine and awareness efforts to distinguish opinions and news, regulation of digital platforms as public utilities, use of governmental antitrust authority to regulate the media, rules to protect media users from having their personal information used in ways that invade privacy and distort the news they receive, payment by platforms for intellectual property of news organizations, regulation of fraud, and robust funding of public media and media education. Reforms along these lines are not simply plausible ideas; they represent the kinds of initiatives needed if the First Amendment guarantee of freedom of the press can hold meaning in the twenty-first century. This book will make the case for the need for change, the legal basis for change, and the specific forms that policy initiatives could take to remedy the failures of the contemporary ecosystem of news, and it will explore ways to navigate potential constitutional barriers to such reforms.

Some argue that the U.S. Constitution has no bearing on the situation. Others go further, adding to the usual obstacles to reform—opposition from industry and difficulties enacting and enforcing new rules—an emerging, aggressively libertarian view of the First Amendment. Because the Supreme Court views the essential trip wire for First Amendment review to be actual governmental action affecting speech, the activities of private digital platforms and private media companies seem beyond the reach of reforms. The Constitution might be interpreted to prevent reforms that touch on or shape the worlds of speech, news, and media. Moreover, by legislation and judicial interpretation, digital platforms enjoy protection from even the limited liabilities for fraud and defamation applied to newspapers and broadcasters.

The forces influencing news production and distribution are intense, and the prevailing legal framework seems unavailing. Yet the freedom of the press defended by the First Amendment of the U.S. Constitution assumes the existence and durability of a private industry. This book proceeds with the argument that initiatives by the government and by private sector actors are not only permitted but required as transformations in technology, economics, and communications jeopardize the production and distribution of and trust in news that are essential in a democratic society. Any contrary view of the Constitution imperils the very system of government it establishes.

Recognizing fundamental changes is often difficult for those who live through them. Yet the decline of the news media is accelerating. The familiar pattern of economic disruption that brought the telegraph, radio, television, and cable complemented but did not destroy the investment in newsgathering and production and the mechanisms for vetting material. The disruption wrought by the internet and the platform companies is not generating a sustainable alternative, and constitutional democracy itself is in the balance.[15] Yet it can be hard to imagine what comes next based only on extending what is current.[16]

Nobel Prize–winning economist Amartya Sen once observed that no substantial famine has ever occurred in a functioning democracy with regular elections, opposition parties, basic freedom of speech, and a relatively free media, even when the country is very poor and in a seriously adverse food situation.[17] In the United States, dangers to freedoms of speech and of the press in the past came from direct and indirect government suppression, such as in 1798 with the Sedition Act, during the Civil War and World War I, and during the Red Scare of the 1950s.[18] In contrast, the current challenges arise from the very digital communication systems that endanger the gathering, reporting, and receipt of news.

Tools of technology, business, and regulation could significantly change the situation. This book describes the current news ecosystem in the United States and the trends that now jeopardize reliable, accessible news for localities as well as for national and global affairs. It then turns to historical developments showing repeated periods of disruption and innovation and also long-standing government involvement in shaping and influencing the news industry. This historical insight provides the basis for the argument that the First Amendment does not forbid government involvement designed to strengthen the viability and reliability of newsgathering and distribution; instead, for more than two hundred years, the First Amendment has coexisted with the aid and activity of government, shaping enterprises for generating and sharing news. An even more ambitious interpretation would treat the First Amendment as mandating such efforts, as an informed citizenry is presumed by and essential to an operating democracy. What is not in doubt is the severe jeopardy in the United States for individuals to access information and the role of journalism in checking falsehoods, making governments accountable, and exposing corruption and other abuses of power. The twelve proposals that close this book represent potential avenues for change. Even if no single one of them alone would fix the current problems, the ideas advanced here suggest steps that can and should be taken.

The global COVID-19 pandemic brought not only great suffering but also crucial lessons. Among those lessons are the essential role that reliable news plays in the age of social media, and the inadequacy of purely private sources of news. The marketplace largely failed to produce and distribute reliable news to everyone who needed it. The pandemic has exposed to view the fragile economics of newspapers and the loss of any enterprise devoted to local newsgathering and distribution in growing numbers of American communities.[19] Does the United States Constitution pose a barrier to fixing the news industry? Or does it provide resources for doing so? What options hold promise for a vibrant, healthy ecosystem of news?

CHAPTER ONE

. . .

News Deserts, Echo Chambers, Algorithmic Editors, and the Siren Call of Revenues

HOW MANY PEOPLE now get news from hard copy newspapers? How many from radio or television? How many from websites? Social media? Reuters Institute found that two-thirds of those surveyed in twenty-six countries use social media, and more people find their news through an online algorithm than through human editors; only one in ten of those surveyed pay anything for online news.[1] People massively rely on smartphone apps and social media. In 2020, only two out of ten relied on print newspapers, and only 5 percent of those ages eighteen through twenty-nine got news from print newspapers. Both during crises—such as the COVID-19 pandemic—and during more ordinary times, social media provides much-desired immediacy and opportunities to participate. Yet the pandemic contributed to further layoffs and reductions of news media while triggering misinformation and conspiracy theories on social media.[2] Local news operations are especially strained and are increasingly laying off staff or even closing up.[3] Newspapers are rapidly shedding staff and shrinking coverage. More than 100 newspapers have shifted from daily to weekly issues, and between 2004 and 2014, at least 664 newspapers shut down.[4] Newspapers reduced employment by 47 percent between 2008 and 2018; largely driven by newspaper staff reductions, overall news journalist employment in the United States dropped by 23 percent between 2008 and 2019.[5] In one month alone, during the COVID-19 shutdowns, more than 30,000 staff at news media organizations faced layoffs, furloughs, or pay reductions.[6] By 2019, over 65 million Americans lived in counties with only one local newspaper, or none at all.[7]

The diminishing presence of local news coverage is especially ironic given that local news tends to be the most trusted.[8] It is also often the most urgently needed. Residents need local news in particular to deal with a public health crisis, to find out about local political candidates, and to know what is happening in their communities. No reporters are assigned to cover the courts of New York's Queens County, which has 2.3 million residents and 200,000 criminal cases each year.[9] Accountability of local governments suffers without the watchdog of local media asking questions and looking more than occasionally at what is going on. When a 121-year-old local newspaper called the *Warroad Pioneer* from a small Minnesota town closed, the community lost coverage of local obituaries, scandal in the county commission, budget crisis in the school district, and high school sports.[10] Mergers of newspaper chains produce more closures and reductions in local news.

Such potentially big shifts in the news ecosystem arise with three trends: (1) corporate investors giving greater priority to financial returns than to quality journalism or maintaining particular local news outlets; (2) news outlet ownership by wealthy individual investors who may support independent journalism or may pursue their own ideological projects; and (3) the shift of advertising dollars to online media platforms that harvest user data. Taken together, the trends help to explain shrinking investments in local news and in professional and specialized journalists. The new owners of newspapers and big digital platforms can choose not to invest in news production or local government accountability—such as reporting on the excessive reliance on fines and fees imposed on people caught up in the justice system in Ferguson, Missouri, where the police shooting of Michael Brown triggered racial protests across the country.[11] It turns out that Ferguson had no daily newspaper, no news blog focused on local government, no community radio station, and no local public access television.[12] Social media may spread the news of police shootings, but it does not provide investigative journalism to expose governmental failures, including creating a structural conflict as courts and agencies supposed to administer justice rely on imposing fines and fees for their own financing, or the piling of fines and fees on poor people in places like Ferguson, Missouri. The crisis of dangerous lead levels in Flint, Michigan's drinking water came to light through local news—but how many other communities have similar problems without journalists exposing the situation?

What is happening to newspapers? What do new owners and consolidation augur for the gathering, production, and distribution of news? How do digital platforms and social media companies affect "legacy" news media and the information that people get and need? What are the effects of these developments on individuals, communities, and the nation?

DECLINING NEWSPAPERS

These trends are departures from better times for news media. For four decades after World War II, mainstream journalism reflected a mission of nonideological reporting about politics, foreign affairs, business, and entertainment.[13] Major broadcast networks helped to unify the country and provide criticisms of government excesses. The journalism of the time was not perfect and always included a range in quality and political slants; still, an ideal of objectivity grew and took hold in professional journalism during the twentieth century.[14] In the past, people generally could easily find reports of local news but did not have immediate access to reports from news media around the world. Now, though, much has changed.

Today in the United States, as journalist Robert Kaiser reports, "the great institutions on which we have depended for news of the world around us may not survive."[15] Coverage of international affairs, even from highly visible national news shops, necessarily diminishes without reporters based around the world. Declining circulation, loss of advertising revenues, and diminishing profits both reflect and fuel the reluctance of many people to pay for news. The introduction of Craigslist and other online sources slashed newspaper revenues from classified ads. Social media sites expanded the lure of the online, and the advertising base of traditional media plummeted. Accessing social media 24/7 without paying for it with money, yet not comprehending how they pay with their data, people are migrating to social media to save money and to save time. Social media platforms take the advertising dollars, and filter and distribute news based on data about what each individual user has liked in the past. The technology allows the speedy spread of eye-catching material, and enables "friends" to rapidly and costlessly share arresting material—including misinformation. Because the platform companies make money by splitting revenues with "influencers" who draw views and by selling ads based on numbers of viewers, the platforms reward conspiracy theorists and spread misinformation, even as some celebrities try to use their social media followings to correct falsehoods.[16]

These trends, though, are complex. Different dynamics are at work for local communities compared with large cities and national markets, and for different people, especially when sorted by age, race, and educational level. But in most areas, consolidation of ownership and cost cutting has diminished coverage of local news.[17] For local news, the problem is profound and structural. A limited number of enterprises will succeed online because both the business model and the networking effects depend upon aggregating readers/users.[18] A consistent focus on the news of a particular locality will never be able to aggregate at the levels of those operating nationally and internationally.

Major newspaper chains have declared bankruptcy, and revenue declines continue. Across the nation, the number of newspaper employees has dropped from 71,000 in 2008 to 38,000 in 2018, and the COVID-19 pandemic that began in 2020 only accelerated this trend.[19] Over the past twenty years, newspapers across the country have lost nearly 40 percent of their daily circulation, and in the past ten years, newspaper advertising revenues decreased 63 percent. However, at least since the election of President Donald Trump, new subscribers to the *New York Times* and *Washington Post* brought those top papers to record numbers and sustaining revenues.[20] As of 2017, 25 percent of those surveyed in the United States say they want to help fund journalism.[21] But almost 60 percent of newspaper jobs in the United States vanished over the span of twenty-six years.[22] News jobs are disappearing across the industry, including in new online ventures, broadcasting, and cable, and especially newspapers.

When newspapers disappear, so do the tent poles enabling local community connections. As consolidation grows also in ownership of broadcasting—with pre-internet, outmoded rules limiting "cross-ownership" across communications businesses—diversity and local engagement diminish.[23] News outlets with a local focus bind communities together by reporting on events, arts, health concerns, opportunities for political engagement, and entertainment. The existence and operations of larger media organizations support smaller outlets that rely on their products. A vital media ecosystem needs reliable reporting not just on a national level but also on metropolitan and neighborhood levels. Radio, television, cable, and internet users feed off major newspapers for their commentaries; all of that is in jeopardy when a metropolitan newspaper shrinks or shutters.[24] Collaborations and shared ownership within geographic communities that once seemed anticompetitive may now be vital in addressing the failures of national conglomerates to focus on local news. A national outlet does not give the local information needed by domestic violence survivors, the results of high school sports events, or the players involved in a dispute over development and neighborhood land use.

Smaller newspapers face steep declines in readers and revenues, with many merging or selling to chains or private equity investors pursuing economic returns through cost reductions and restructuring.[25] As papers such as the *Rocky Mountain News* close, and others reduce the frequency of issues from daily to weekly, local news updates are less available, and because of the relatively small numbers of individuals affected by any particular closing, internet solutions are not likely. In earlier times, concentrated ownership by Knight Ridder and Times Mirror elevated the quality of many local news outlets, but even they ended up making serious cuts before selling remaining assets to new buyers.[26] Investment-focused owners and chains have been buying up local papers, producing unprecedented levels of consolidation.[27] The ten largest chains have doubled their reach in recent years, and the number of daily newspapers continues to decline.[28] Only twenty-five companies owned two-thirds of the country's daily newspapers in 2018.[29] Concentrated media ownership may be more an effect than a cause and would not itself undermine the new industry if owners committed to the enterprise of producing quality journalism rather than stripping assets for profits. But owners can and do close individual papers that do not make the profit they seek, reduce reporting about local news in favor of more generic material, and focus on stories that will "trend" rather than provide the kind of news that equips people to govern themselves. The publicly traded companies holding many newspapers need to turn profits for shareholders, even if that means sacrificing journalistic values.[30] Smaller staffs means fewer resources for journalists who are specialists in fields such as science and the environment, greater reliance on press releases, and diminished investigative journalism.

In some cases, the result is essentially "ghost papers," existing in name but using content generated elsewhere with no specific connection to the community they are supposed to serve. People need more news about economic, political, and governmental matters to navigate health care coverage; to deal with credit cards and mortgages; to oversee schooling for their own and other people's children; and to understand local recycling rules, large environmental risks, and a host of other issues. If anything, these needs are growing just when likely outlets may be less able to generate and distribute effective information.

Some nonprofit news organizations are emerging to address the declining presence of prior news outlets. Experiments are especially necessary to reach younger audiences who may increasingly discount or ignore traditional news publishers.[31] New start-ups such as Spotlight PA, a partnership of local Pennsylvania papers for investigative reporting, may fill some of the voids, reporting on governmental and business actions or monitoring police conduct in a particular neighborhood, but these efforts, whether supported by philanthropy, volunteerism, or venture funding, have not found a sustainable path.[32] The private sector simply may not be able to generate sufficient funding for the kind of reporting that holds local governments accountable. The entire business model of newspapers—print or digital—is "very much in free fall."[33] As people grow reluctant to pay for news that is posted for free on the internet, ad revenues migrate to digital companies, and digital companies themselves invest little in news gathering, editing, and reporting.

NEW OWNERS

A few prestigious newspapers have found wealthy individual investors. Jeff Bezos purchased the *Washington Post*, Patrick Soon-Shiang purchased the *L.A. Times* and the *San Diego Union-Tribune*, and Laurene Powell Jobs's Emerson Collective owns a majority interest in the *Atlantic*. Some investors may be philanthropically minded, leaving editorial decisions to professionals, but others may be seeking to influence the political tilt of the news or to change it in other ways. The Mercer family's control of Breitbart News Network, Charles and David Koch's pursuit of media ownership, and Rupert Murdoch's transformation of a family media business into an empire on three continents are examples of efforts to use wealth to advance particular ideologies through media.[34] (Rather than buy an outlet, another billionaire—Peter Thiel—used his resources to destroy a media outlet by financing a series of lawsuits against Gawker Media, which led to its bankruptcy; Gawker eventually ceased operations.)[35] Most mass media remains held by private companies or publicly traded corporations; purchases by high-wealth individuals and by private equity funds are notable developments, affording greater power by a few individuals over the affected news operations.

With broadcasting and cable, mergers and consolidation similarly risk diminishing local news and reducing diversity of opinions and viewpoints. For example, Sinclair is a company that owns 173 television stations, through which it spreads right-wing political perspectives.[36] Sinclair planned to purchase forty-two more stations, allowing it to reach three-quarters of American households, but the Federal Communications Commission (FCC) has, for now, blocked the effort and found—by a vote of 3–2—a variety of unrelated legal violations.[37] Such concentrated ownership displaces local control of media and shifts editorial decisions to people without a stake in particular local communities. Many local news shows look just like local shows in other parts of the country because stations now borrow segments from other stations owned by the same company. Local television news turns to weather, traffic, crime, sports, banter, and entertainment news, and national broadcast news networks have cut costs, staff, and coverage.

For better or worse, traditional television is losing viewers—especially younger ones—to streaming services and other digital alternatives. People between the ages of thirteen and twenty-five watch less than thirteen hours of television a week, which is 44 percent less than five years ago for people in the same age group.[38] "Cord-cutting"—terminating or not starting cable subscription—is the trend, as younger people especially turn to streaming services. This development affects even traditional media programming. Since the rise of cable and the internet, broadcast news has shifted to more entertaining and profit-conscious news programming, and even newsmagazines have shifted to prefer emotional stories over factual investigations. Cable and digital communications allow subsets of the population to connect, offering avenues for more diversity in programming and voices, but this also encourages "narrowcasting"—aiming for slices of the community rather than trying to reach everyone. Meanwhile, older people reminisce about the golden age of television news, and remember moments such as when Edward R. Murrow did war reporting on the scene and met the Red Scare led by Senator Joseph McCarthy with courage, or when Walter Cronkite narrated moon launches, the assassination of presidents and other political and civic leaders, and the Vietnam War.[39]

DIGITAL PLATFORMS

New competitors to the concentrated media industry disrupted their business models with popular streaming services, leading networks to play catch-up, while data and digital platforms, including Amazon, YouTube, and Netflix, changed the way vast numbers of people find news. In an example of what law professor Frank Pasquale calls "the black box society," data platforms customize people's access to news (as well as sports, entertainment, and other content) without even consulting them. Instead of offering clear choices, digital platforms bury decisions that affect people in the architecture of their sites, relying on analyses of computer data usage that is opaque to users.[40] As one service advertised in its launch, "The feature delivers a way to browse and discover news from publishers worldwide, and introduces a personalized newscast—through a 'filter bubble'—that adapts to your interests based on what programming you watch and skip, among other things."[41] A number of these services sample content from elsewhere, while others generate their own stories. Some commentators maintain that these new services will cover multiple sides of an issue, and one observer urged people "to be careful not to create your own echo chamber in which you only ever hear opinions you agree with."[42] Even though people do not only read what they already believe, this kind of expert advice is beside the point if people do not know why they are seeing what they see and do not have the tools they would need to encounter anything else.

Attention and money are now concentrated on a few digital companies. Facebook, for example, had 1.6 billion participants around the globe in 2016; by 2019, the company reported 2.5 billion active monthly users.[43] Its number of users has surpassed the number of people in the world's most populous nation.[44] Facebook's recent effort to highlight social content (promoting posts from family and friends over content from publishers in the personalized "newsfeed") risks further deemphasizing news, and especially local news.[45] The shift of dollars and attention to social media platforms carries further risks. As people adopt ad-blocking technologies, the platforms move to "native advertising," blending ads with content.[46] Social media platforms draw attention and advertising revenues away from traditional media while using and selling data about each user's clicks and engagement. The effect is to make the user into the product and potentially provide easy vehicles for those who profit from increasing social division, fomenting hatred, and undermining democracy.[47]

Leaders at Facebook and Google have stressed that, as tech companies, they are not in the business of journalism. They rely on algorithms, sometimes overriding human editorial decisions, to select what people see.[48] They focus on keeping consumers' attention, not on covering the news. Sheryl Sandberg, the chief operating officer of Facebook, explained: "We're very different from a media company. . . . At our heart we're a tech company. We hire engineers. We don't hire reporters. No one is a journalist. We don't cover the news."[49] And yet, more and more people get their news from social media, through links and forwarded posts—with each act of sharing increasing a post's visibility to others. British reporter Emily Bell noted, "Social media hasn't just swallowed journalism, it has swallowed everything. It has swallowed political campaigns, banking systems, personal histories, the leisure industry, retail, even government and security."[50] Judgments once made by a variety of people with diverse aspirations are now made by profit-maximizing algorithms seeking to capture the largest number of "eyeballs" and advertising dollars. Algorithms deploying machine learning and data about individuals determine what each user receives on their Facebook feed, Twitter timeline, and YouTube home page—and the big platforms use them to amplify emotions in order to maximize attention.[51] The capacity to make editorial judgments remains, as revealed by many platforms' rapid removal of disinformation about the coronavirus. But social media and tech platforms do not often use those capacities. And when they do use them, there is no assurance that the guiding norms will advance free expression, safety, or any other civic values, nor that the users and the public will understand what values and procedures are at work. Moreover, some worry that user engagement drops if the platform adopts a norm such as improving the quality of the news content highlighted by an algorithm.[52]

Trade-offs between speech and truth, between scale and safety, and between profits and democracy may be unavoidable; should those trade-offs, though, be made in secrecy by a few private internet companies? Journalist Evan Osnos concludes that Facebook founder Mark Zuckerberg "is at peace with his trade-offs": "Between speech and truth, he chose speech. Between speed and perfection, he chose speed. Between scale and safety, he chose scale."[53] But, continues Osnos, "like it or not, Zuckerberg is a gatekeeper. The era when Facebook could learn by doing, and fix the mistakes later, is over. The costs are too high, and idealism is not a defense against negligence."

Ostensibly neutral digital platforms are easily manipulated by propagandists and extremists who use search optimizing and digital clicks for their own ends, while offering revenues to Facebook and Google. Facebook, for example, has become a tool of choice for Rodrigo Duterte, the autocratic president of the Philippines.[54] In the Philippines, 97 percent of people have Facebook. Duterte's support from Facebook started with training sessions the company provided for presidential candidates and continued with "white-glove" services upon his election. He and his supporters deployed fake accounts, aggressive messages and insults, threats of violence, and fraudulent endorsements, creating the illusion of support for his regime. He then used Facebook to stream his inauguration after he banned all independent media in the Philippines. Facebook, in turn, has entered into a partnership to lay undersea cables to support users in the Philippines and allowed critics of Duterte to be removed from Facebook. Similar special services for politicians could be offered in other countries, including in the United States.

While leaders of digital services have claimed they are passive intermediaries treating everyone the same, critics charge that the tools and designs at work enable abuses.[55] Such charges have not produced successful verdicts because data companies like Facebook, Twitter, and Google have avoided civil liability. Some allege that these platforms and their tools assist terrorist groups such as Hamas; again, the companies claim that they are not responsible for the content on their platforms.[56] Leaders of digital companies have pretended that their platforms make no editorial choices for which they should be responsible. Yet tech platforms fundamentally shape content and intervene to influence what people see through moderation decisions, deletions, highlighting or submerging content, and granting privileged access; these activities are in fact the business in which the tech platforms specialize.[57]

With the stay-at-home orders imposed in the wake of COVID-19, Facebook put its 15,000 content moderators (contractors who worked in offices scattered around the globe) on paid leave and turned even more dramatically to artificial intelligence tools for content moderation.[58] Twitter showed its ability to moderate content by labeling potentially harmful or misleading information related to the coronavirus with a tag reading "Get the facts about COVID-19" and links to a page curated by Twitter or created by an external source.[59] These measures have not, however, been pursued with the same seriousness and resources with regard to hate speech, terrorist speech, conspiracy theories, or "fake news."[60] Instead, the burden of checking facts has largely remained with users, who are often ill-equipped to sort out fact from myths or manufactured material.[61]

And the platform companies make no commitment to the principles of independent journalism.[62]

Whether or not the tech platform companies are aware of what is happening, the tools of the digital companies are easily used to spread misinformation and fraudulent content. When special counsel Robert Mueller indicted thirteen Russians for disrupting the 2016 United States presidential election through Facebook and other digital media, he effectively torpedoed denials by Facebook executives about the platform's role in election-season misinformation and propaganda.[63] Russian provocateurs, knowledgeable about social media, used widely available technological tools, perhaps including some not known to the companies themselves.[64] Even if such content might be protected by the First Amendment, the norms of professional journalists would test and filter out misinformation and propaganda. How much does the insulation from civil liability that is presently afforded to digital platforms lead to insufficient precautions against such exploitation and misuse?

The way digital platforms are immune from civil liability differs from the treatment accorded to newspapers and broadcasters, which can be held civilly liable for defamation, false information, threats, sexually explicit material involving minors, and racially discriminatory housing ads posted by users.[65] In contrast to legacy media, digital platforms have protection under Section 230 of the Communications Decency Act.[66] Though this immunity was created to enable the innovation and expansion of digital platforms, arguments for revising this section of the act are increasingly drawing attention and support from people across the political spectrum.[67] But as now required with an amended Section 230 accountability for failing to warn about a known online sexual predator or for hosting a site that matches potential roommates in a racially discriminatory manner, it has not deterred platform companies from providing and expanding their services.[68]

The role played by digital platforms in misinformation and propaganda is significant because of the sheer number of people who engage with those platforms. The use of algorithms accelerates the spread of materials that attract attention. Readers are often vulnerable to hoaxes and abuses enabled by "dark posts"—ads that are invisible to all but those targeted and that do not reveal who paid for or is behind them.[69] Oxford University scholars have studied and critiqued this "computational propaganda."[70] Activists can use digital media to nudge voter turnout and target individual voters. "Clickbait"—arresting headlines and attention-drawing ads—enables a surprising amount of disinformation without the checks that counterspeech (fighting problematic speech not with censorship but with more speech) and investigation can provide.[71] Filter bubbles, critics charge, isolate individuals in a stream of messages that match their prior views. So do unscrupulous campaigns of division that include attacks on the media. Even if only small subcommunities of people echo one another without challenge, that becomes a problem not just for individuals but also for society and democratic processes.[72] Digital platforms can deploy "digital gerrymandering," selectively presenting information to serve interests unknown to recipients and undisclosed to the world.[73] Although Facebook has added staff to police hate speech and take down fake accounts, no one thinks these efforts work well.[74] Investigations into past and present risks of chaos and misinformation continue.

Some argue that the concerns are overstated, while others point to deeply worrisome patterns. One study based on uses of Facebook during the 2016 U.S. presidential race argues that any effect of online exposure on polarization is modest.[75] Another study, again about social media during the 2016 presidential election, maps a distinct difference between right-leaning sites, cable networks, and broadcast media and left-leaning ones: the right-leaning outlets appear siloed off from other media and news circulation, allowing rumors to spread without correction or contradiction, while left-leaning sites are in dialogue with and checked by other sources, curbing disinformation.[76] Yet still another study shows that false information spreads more rapidly than verified information from reliable sources.[77] Nonetheless, an exposé documentary about Facebook, identifying its role as conduit for hateful materials, generated vociferous responses by the company.[78]

EFFECTS

Failing business models for newspapers, new owners with varied agendas, digital platforms disrupting communications and drawing advertising revenues (including from classified ads) away from legacy media and especially away from local news, shrinking viewership for broadcast news: these and related trends contribute to the crisis in journalism and the news business in the United States. The effects on individuals, communities, the nation, and government accountability as well as democratic governance are multiple and cascading, as documented by a comprehensive federal government study.[79]

When newspapers close, local government becomes more expensive to taxpayers, no doubt reflecting the absence of monitoring of government salaries, debt, and other expenses.[80] Decreased awareness of local issues reduces voter turnout and engagement in civic affairs.[81] The for-profit model of newspapers supported largely by advertising does not work when cheaper, targeted online ads replace print media ads. Recent estimates indicate that 89 percent of online advertising dollars go to Google or Facebook, and 60–70 percent of all advertising revenues go to internet companies.[82] As advertisers redirect their dollars away from undifferentiated print media and toward data-driven, targeted online advertising, online tech companies benefit and traditional media and the news industry suffer.[83]

The pattern is even more pronounced when it comes to local news markets.[84]

Unbundling is another trend that has hurt traditional media. In the past, even when hard-core news struggled for readers, newspapers and broadcasters could help pay for it through cross-subsidies offered by bundling content: people interested in style, crossword puzzles, and horoscopes would help pay for reports on politics, science, and sports. Now, though, people can get their updates about sports or weather apart from political news, further decreasing the cross-subsidies once available to newspapers and other mass media.

The negative effects of digital media on elections exacerbate the decline in people's trust in media that is already underway.[85] According to a 2017 poll, nearly half of registered United States voters believe major news organizations made up stories about Donald Trump.[86] Professional journalism, messages from your cousin, or messages from a Macedonian adolescent paid to design arresting ads can seem equal in a world without editors vetting stories. Individuals presenting themselves anonymously online may be demonstrating the benefits of free speech but may also accelerate the destruction of basic norms of civility and honesty; the same is true for ads that give no clue about who funded them. Social media algorithms that determine what is distributed to whom are not visible to anyone outside the companies, and it can take quite a while before the actual patterns of distribution are apparent.

Contributing to the growing distrust of news and media is the absence of any clear regulatory guidance due to the political deadlock in the Federal Election Commission, the government agency charged with regulating election-related speech.[87] In early 2018 YouTube promoted a conspiratorial video accusing one of the survivors of the mass shooting that had occurred at Florida's Marjory Stoneman Douglas High School a few weeks earlier of being an actor who did not attend the school. YouTube later explained that it misclassified the video originating originated from a local television outlet; meanwhile, the video was viewed more than 200,000 times, labeled as "trending," and accelerated in its distribution before complaints led to its removal.[88] YouTube's algorithms tend to recommend channels advancing conspiracy theories and falsehoods, even to people who have never shown an interest in such content. Frequently changing rules about how much power internet providers and data service companies can exercise over what users see or know about only adds further complexity and confusion.[89]

Surveys show that a large and growing number of Americans see news stories as faulty and the professional press as unwilling to admit mistakes or correct biases.[90] President Trump made "fake news" a popular phrase, and disagreements over the meaning of the term just amplify the doubt and distrust of providers of news.[91] Competing for shrinking audiences, even mainstream media increasingly stress sensational headlines or human-interest stories. At the same time, technological and economic disruptions alter how news is gathered, edited, accessed, distributed, and financed. Tech companies, guided by psychologists, organize the attention of users while traditional journalists are just beginning to learn about the neuroscience of creating receptivity for reasonable but skeptical audiences. These patterns contribute to and reflect the diminishing role of evidence and analysis in the United States.[92]

In some ways, the current disruptions in media and news echo past technological and economic changes that have radically altered the vehicles for collecting and distributing news in the United States. Big data platforms draw patrons and advertising away from older media. Perhaps older companies and media will find ways to adapt by collaborating with new enterprises and by focusing on their distinctive strengths; just as newspapers moved to more in-depth analysis after the development of 24/7 broadcast news, all media will change in response to and in collaboration with data platforms on the internet.[93] Federal and state governments play catch-up, pursuing regulation after developments emerge, and chiefly rely on competition among private sector companies to check bad practices. Technological advances allow government officials (such as Presidents Franklin Delano Roosevelt and Donald Trump) and, indeed, anyone with a message to spread to go over the heads of professional journalists and send those messages directly to the public (through radio, in FDR's case, or Twitter, in Trump's). And the roles of professional journalists have tended to shift from reporting events to analysis. In both scope and consequences (which are still unfolding), the transformations of communications and related industries in light of the digital revolution are more akin to the changes ushered in with the book than to the impact of broadcasting and cable television.[94]

The current shift differs precisely because of the availability and practices of new media and digital tools. Digital networks, unlike the telegraph, radio, or television, do not need the newspaper to travel the "last mile" to the reader, and they allow readers to communicate across the networks too. As one journalist put it, "Google stole the delivery trucks and Amazon stole the newsstand."[95] The lag time of a single day once was enough to allow one newspaper an advantage in breaking a story, but now internet platforms, as well as broadcast and cable, can transmit any story reported by a newspaper immediately—with most of the profit not going to the news outlet that initially reported the news. Through the network of networks that composes the internet, one-to-many and many-to-one communications are easy and can bypass the newspaper, publisher, or broadcaster that used to select, edit, and vet news. Anyone with access to an email account, mobile phone, or social network can not only receive but also send out information. These channels are global and "distributed," meaning the components are spread widely and coordinated through networks, rather than through hierarchies or a central coordinator or chain of command.[96] And now advertisers can bypass newspapers and other intermediaries, reaching customers directly online and gathering data about them at the same time.

Because digital mega-companies, such as Google, Amazon, Facebook, and Twitter, generally do not gather, edit, or produce news stories, their dominance of the news business does not strengthen or even preserve reliable news. Indeed, the digital companies free ride on the content generated or conveyed through social media by users, and also use data analysis, behavioral nudges, and social marketing to gain customers and to sell data about users. Unlike newspapers and broadcasters, the digital companies do not need to invest in gathering or assessing news or in serving local communities without large populations. And extreme radio and cable providers pushing outrage add to the general weakening of legacy media; professional journalism has little role in controlling the limits of what is acceptable to say and what is believable.

Even the huge digital companies are only partially in control of the transformed world determining the shape, price, and quality of news. You send and receive news and other communications by connecting to a telecommunications service provider, which translates the text message sent or received into electronic signals, transmitted through the network as packets of data through an internet service provider, such as Verizon, AT&T, or Comcast, and ultimately this chain translates a message back into text when it reaches another device.[97] If internet service providers do not treat all data packets the same way, individuals or organizations that have more money or influence or that are favored by a controlling government will be able to negotiate better speed and service.[98] When you use the internet, you are using a device—a phone, tablet, or laptop computer—that is linked to hardware, such as an Ethernet network card or a modem, connected to an electronic cable, wireless transmission of radio waves, or a beam of light sent down an optical fiber.[99] Private companies compete in assembling this network of networks we call the internet. The businesses and architecture of each of these elements affect whether and how people receive or create news and other information. Held by a few dominant companies, each of these elements raises new concerns about concentrated and unaccountable power.[100] Despite early hopes that internet communications would create more opportunities, powerful entities such as Facebook have even more concentrated power than prior companies that dominated communications.[101] These companies—and the advertisers using their channels—can bypass newspapers and broadcasters as vehicles for reaching audiences.

Further complicating the situation is the sheer complexity and lack of transparency of the internet and digital companies.[102] Robert Mueller's indictments give a clue about the vulnerability of Facebook tools to manipulation and deceit, but it is difficult for most people to see or understand fake accounts, micro-targeted ads, or devices such as "dark posts," which are not "published" on the page but appear only to the user.[103] Users confuse items labeled as "sponsored content" with vetted news stories, according to studies.[104] Innovations blurring the distinction between ads and news make the problem much worse.

What other techniques are available for misuse? Mathematical formulas—algorithms—are supposed to connect people with content they are looking for and would like. What if a digital platform adjusts the math "so that only posts that get a disproportionate amount of engagement (likes, clicks, comments, shares) will be seen by a lot of people—regardless of whether those people are fans or friends"?[105] Traffic is directed by an unseen traffic cop. Someone who likes updates from friends may not get many of them because they are not circulating among or preferred by the large numbers of people valued by the algorithm. Companies and people who are not getting their messages delivered as much as they would like can—and do—pay for better placements.[106]

This is the technology that enables predatory messages, such as ads for fraudulent educational opportunities, targeting vulnerable people with false or misleading information.[107] Facebook in 2002 altered the newsfeeds of two million politically engaged people, who were sent a higher proportion of hard news instead of cat videos and the like; when their friends shared a news story, it showed up high on their feed without revealing the mathematical tweak behind this phenomenon. Unlike choices by Fox News or MSNBC editors to highlight one story over another, these algorithmic adjustments are invisible to viewers and thus elude comment or criticism while adopting psychological ploys to keep users hooked. Apparently, during the 2016 election season, fake stories, especially pro-Trump or anti-Clinton, attracted engagement more than legitimate news did; some of those fake stories generated income for teenagers in Macedonia who created websites and messages repeating such hyperpartisan content.[108]

Both carefully designed and reflecting unconscious biases, algorithms govern the work of digital platforms as they collect and use vast amounts of information about individual users. And the data guiding the distribution of content may be faulty. People are known to lie or boast on Facebook, for example, but the algorithms sweep up these reports along with truthful ones.[109] People using digital platforms such as Facebook and Google do not have a window into the choices made in the design of platforms, and yet those choices select, suppress, push, and censor.[110] This visibility problem exacerbates the filter bubble—the intellectual isolation that can result when algorithms select what users should see based on predictions about what they would like.

Whether social media causes or reinforces separate communities is less important than how social media contributes to a world in which some people receive dominant messages that COVID-19 is a severe health hazard, others receive waves of messages announcing it is a hoax, and some receive both. The spread of a conspiracy theory linking 5G cellular networks to the pandemic spread quickly around the world even though the newspaper that originally published the article quoting someone making that claim retracted the article within hours.[111] Boosted by right-wing website InfoWars and by English former soccer player David Icke, the conspiracy theory spread widely.[112] There have always been conspiracy theorists and liars, but now internet companies give them frictionless global distribution for no financial cost. Researchers found that 48 percent of American adults reported encountering "at least some news and information about COVID-19 that seemed completely made up, with 12 percent saying they have seen a lot of it and 35 percent saying they have seen some." For example, nearly 30 percent of Americans believe that COVID-19 was made in a lab.[113] The eventual removal of conspiracy-related posts by Icke and other individuals does not undo the spread of the misinformation. The tech platforms are not responsive to the many social media users who want to ban false information about the virus conveyed by public officials and also want misinformation about the virus removed from all social media platforms (although there are currently no mechanisms to do so).[114] While the pandemic and shelter-at-home practices contributed to boosts in the use of both conventional news media and social platforms, 50 percent of Americans in communities reporting COVID-19 cases had access to little or no local news media and hence faced grave limits on local news and information about the availability of local COVID-19 testing.[115] Despite increased interest in local news, local newspapers laid off or furloughed employees as business declined during the coronavirus outbreak in the United

States.[116] Americans who mainly receive their news through social media were the least likely to follow coverage of the pandemic and the most likely to report encountering unreliable material.[117] Social media sites, once reluctant to remove posts, worked actively to eliminate misinformation about COVID-19, but much of the information remains, including some presented in documents that look official.[118] The World Health Organization has called the rapid spread of misinformation about the virus an "infodemic," a serious situation requiring an aggressive response.[119]

Facebook and Fox News both contribute to the spread of misinformation, and tracking the problem exposes the deeper vulnerabilities in the media ecosystem.[120] Dominating the circulation of news, rumor, and misinformation globally, American-based companies regulate content, allow companies and governments to censor and monitor individuals, and make the rules governing what users can see.[121] Social media amplifies people's prior views and predicted interests, and may contribute to social division and polarization, even before enemies of the United States exploit domestic rifts.[122] At times Russian disinformation even appeared in traditional U.S. journalism outlets, but as former U.S. ambassador to the UN Samantha Power noted, "Russia has keenly exploited our growing reliance on new media—and the absence of real umpires."[123] Russian and Macedonian actors pretending to be Americans are a transborder threat to democracy that is no less serious than cyberhackers' threats to American businesses.[124] Recent research, though, suggests that both greater domestic forces and political asymmetry contribute to polarization and finds that right-wing sites and their users operate apart from the checking function of mainstream media.[125]

Transnational digital platforms daily inject uncertainty, manipulation, and fraud through the open digital architecture housed by dominant private companies that are either unwilling to provide or are incapable of providing reliable messages. A former Facebook executive explained how the company was taken by surprise: "You're so focused on building good stuff . . . you're not sitting there thinking, if we get lucky enough to build this thing and get two and a quarter billion people to use it, then this other bad stuff could happen."[126] And while new communications technologies have long attracted propagandists—the great public relations expert Edward Bernays, who convinced large numbers of American women to smoke cigarettes, noted how shifting news technologies give rise to new periods of propaganda and fakery[127]—what is new is how the current technologies allow the propagandists to hide their tracks from even the most observant critics, while unleashing distractions and distortions on an unprecedented global scale.

Another problem beyond those discussed arises because the internet is vulnerable to "flooding," in which an enormous volume of information drowns out disfavored speech and either discredits mainstream media sources or distracts people from them. Bot attacks boost harassment on the internet.[128] And "internet trolls" post inflammatory, provocative messages to disrupt online communities or badger individuals into withdrawing.[129]

The government was once assumed to be the main threat to the "marketplace of ideas," through punishments or bans on publication, but now the greater danger comes through overwhelming individuals with messages that swamp meaningful communication. Although "more is better" once seemed a sensible approach to freedom of speech, the "more" provided by digital resources may destroy professional journalism, undermine public confidence in information, and negatively affect the provision and absorption of information needed for self-government. The current situation differs from prior disruptions because now the very viability of news enterprises, of getting local, regional, national, and international news to people, is under siege.

This is the new ecosystem of news. Serious risks of news deserts; echo chambers; concentrated ownership of newspapers, radio, television, and cable sources of news; a shrinking number of professional journalists; blurring of ads and news; and dominance of digital platform companies: this is what shapes people's encounters with news. As the big digital platforms do little to invest in news creation, the United States is in danger of losing the crucial relationship between the press and democracy—that is, holding officials accountable—as presumed by the authors of the First Amendment.

CHAPTER TWO

. . .

News Production and Distribution in the United States: Private Industry and Government Contributions

BECAUSE OF THE importance of the press and news to democracy, government policies and programs have provided resources and guided new developments from the nation's founding into the present. The federal, state, and local governments within the United States have contributed money and devised laws and regulations to develop the free and private media, consonant with the values underlying the First Amendment. Over two centuries, governments have made public investments in the development of news media. The federal government has occasionally used its power to regulate markets to shield innovative media from competition; at other times it has exerted that power to enforce competition rules in order to promote access and innovation. As repeated waves of technological change and innovation in reaching audiences and financing media challenged prevailing methods of sharing news, government policies shaped, supported, and protected media and promoted development of diversity and quality in content. The transformation of media from printing presses to the internet thus involves both the ingenuity of private enterprise and intentional government policies. Both have been crucial to the operations of freedom of speech and of the press. Although private sector companies and investments are central to the development of media news, government subsidies and regulations have long played influential roles, especially in developing the telegraph, broadcasting, cable, and the internet.

The history of American communications and news businesses requires focus on private enterprise, but for the bulk of American history, governmental involvement has been integral to the structure, financing, and effectiveness of the news industry and media. This critical and ongoing role of government in American media exposes as false any claim that the First Amendment bars government action now. The disruptive dimensions of the digital revolution are distinctive only in the relative passivity of government in attending to effects on markets, quality, and democracy.

PRIVATE ENTERPRISE

Starting with the private enterprise side of the equation, journalism grew from individual printers to party-supported newspapers and then to a mix of small enterprises and large and profitable corporations. Once the province of solely individual printers and small businesses, the private media now includes not just those types of entities but also large publicly traded corporations and investments held by private equity firms, wealthy individuals, and nonprofit organizations. Private enterprises invest in and launch communications satellites and research into more technological communications advances on the horizon.

From the nation's start, the informed electorate and accountable democratic republic it serves hinged on the vitality of private industry. Homegrown broadsheets brought information about the Old World, the Revolution, and the new nation; they expressed partisan views and opposing views; and they printed letters from readers along with materials from other newspapers.[1] The freewheeling, unruly press was part and parcel of the revolutionary spirit launching the country.[2]

The framers of our Constitution and Bill of Rights understood the crucial role of the press. James Madison, for example, saw public opinion as the real sovereign in a free society. But in a large society, public opinion is "less easy to be ascertained, and . . . less difficult to be counterfeited."[3] Accordingly, he argued, freedom of the press would be crucial to ensure "a general intercourse of sentiments," including roads and commerce, "a free press, and particularly a circulation of newspapers through the entire body of the people."[4] These elements are the preconditions for the "republican form of government" guaranteed by the Constitution vesting sovereignty in the people and in their chosen representatives.

During the revolutionary period, freedom of the press, both in fact and as a symbol, played a crucial role in the demands for self-government and governmental accountability. The practices were built on English and Dutch experiences, particularly freedom from government requirements of religious orthodoxy in the press.[5] The landmark Virginia Declaration of Rights stated that "the freedom of the press is one of the greatest bulwarks of liberty and can never be restrained but by despotic governments."[6] Founder John Adams stressed the people's right to knowledge about the character and conduct of those in charge.[7] In one of the events inciting the American Revolution, Great Britain tried to raise revenues through the Stamp Act of 1765 by requiring American colonists to pay a tax on every piece of printed paper—including newspapers and other documents.[8] The protests that followed focused on taxation without representation, and it was also obvious that the tax could destroy American printing businesses. Colonial newspapers, once bland and noncontroversial, began to mobilize opinions against Britain.[9] Patriot printers generated newspapers and pamphlets, including Thomas Paine's *Common Sense*, to present arguments and information about the conflict unfolding between the mother country and the colonies.[10] Thomas Jefferson wrote to his friend the Marquis de Lafayette, "The only security of all is a free press."[11] Journalists continued to print criticisms of politics and officials throughout the Revolution, and leaders celebrated the debate expressed and fostered by the press as key to the development of a new country. The Continental Congress sought support for their cause, in part, by extolling the freedom of the press:

> The importance of this consists, besides the advancement of truth, science, morality, and arts in general, in its diffusion of liberal sentiments on the administration of Government, its ready communication of thoughts between subjects, and its consequential promotion of union among them, whereby oppressive officers are shamed or intimidated into more honorable and just modes of conducting affairs.[12]

Freedom of the press epitomizes liberty for all in a nation founded in government by the people. State constitutions, and then the Bill of Rights amending the United States Constitution, emphasized freedom of speech and of the press. Historian Leonard Levy concluded that for the founders, "freedom of the press had come to mean that the system of popular government could not effectively operate unless the press discharged its obligations to the electorate by judging officeholders and candidates for office."[13]

Newspapers grew, along with the new nation of the United States, and even reached into most small towns, more so than in British North America (now Canada) during the same period. Competition among newspapers increased as political parties grew over the course of the nineteenth century; newspapers became partisan to cultivate readers, and some actually received subsidies from political parties.[14] Some communities without newspapers mobilized to get one by offering credit to a printer, ensuring a sufficient number of subscriptions, or finding a political sponsor. Demand for local papers may have reflected the decentralized nature of the government.[15] By contrast, in France and Britain, publishing and news enterprises were concentrated in the capital cities.

Although enacting the First Amendment was important to the ratifying states, during the nation's first century the constitutional language forbidding Congress from abridging the "freedom of . . . the press" had no judicial enforcement.[16] It remains unclear whether the framers contemplated a right for the people to receive information rather than simply protections of the press against censorship, although the rights of readers and listeners appeared occasionally in their writings.[17] During the colonial and early national history, printer-editors struggled to survive; they learned to print practical advice and anonymous political pamphlets, and they pursued partisan positions that, over, time connected to the emerging political parties.[18] Even in those early days, the press was largely a commercial effort by printer-editors who relied on private markets for sales, advertising, and subscriptions, but the enterprises received government assistance in the form of postal subsidies.

Commercial newspapers searched for ways to appeal to audiences and to secure financial support. Early printers printed not just truths but also rumors or invective. Although not contemplated by the Constitution, political parties emerged soon after the nation's founding, and newspapers claimed partisan loyalties and affiliation while pursuing audiences and financial support.[19] The partisanship could be intense and critical. Nonetheless, key figures reinforced dedication to the free press. Thomas Jefferson did not always welcome press criticism.[20] Yet while he was secretary of state, Jefferson advised President George Washington that "no government ought to be without its censors: & where the press is free, no one ever will."[21] Commercial newspapers, run for profit, relied on private markets for advertising and subscriptions, as well as on the subsidies from the lower postal rates. By 1822, the country with the largest number of newspapers in the world was the United States.[22]

During the 1830s, selling a daily issue of a newspaper cheaply—for as little as one penny—became a successful business model, as even laborers and clerks could afford to buy a paper. As the "penny press" reached out to working-class readers, the press shared more fact-based information, human interest stories, melodrama, and gossip, instead of the opinion-based articles that had dominated elite publications.[23] Weekly and daily newspaper circulation approximately doubled between 1830 and 1840.

As the new country grew during the early nineteenth century, so did newspapers, with printers and papers even in small towns. William Lloyd Garrison apprenticed with a newspaper printer and then launched a paper of his own, which failed, before accepting a post as editor of a newspaper devoted to the crusade against slavery.[24] Free Blacks founded newspapers as well, leading to about forty such newspapers by the time of the Civil War.[25] Steamboats, railroads, and steam-operated printing presses accelerated the transmission of printed news. Circulation boomed, and during the Civil War papers added maps and illustrations to provide information about battles and casualties. After a private company built the first transcontinental telegraph, initiated in 1848 and completed in 1861 with the aid of state and federal subsidies, some newspapers participated in joint ventures, sharing costs of telegraph messages and other news dispatches.[26]

Newspapers solidified a model with specialized roles for publishers, editors, compositors, and people selling ads and subscriptions. After the Civil War, journalism grew even more commercial and more profitable, benefiting from the telephone for newsgathering and from new methods for printing photographic images and cartoons. The number of daily newspapers grew fourfold between 1870 and 1900; circulation increased as well. Reporting on scandals, architectural and technological change, and the mistakes of those in power, newspapers gained influence and profits that owners often plowed back into the enterprise. Magazines too gained readers and built national audiences in part by presenting investigative work by muckrakers and in-depth news analysis alongside fiction and housekeeping tips.[27] By the 1920s, many more issues of newspapers circulated than the number of households in the United States, and 95 percent of Americans read the papers.[28] The technological and economic dimensions of change generated and reflected legal and political action—governmental contributions—that in turn shaped the development of the media ecosystem.

SHAPING THE PRESS THROUGH GOVERNMENT FINANCING, REGULATION, AND TECHNOLOGICAL AID

From the start of the new nation, the government of the United States attended to and influenced the media's gathering and distribution of news. In 1792, Congress exercised its constitutional authority to create a postal service, with the goal and effect of broad and comprehensive influence on communications and distribution of news.[29] The postal system enabled newspapers, books, letters, and pamphlets to circulate even to remote villages. It also enabled newsgathering and news sharing—Congress gave newspapers discounted postal rates and authorized the free exchange of newspapers with other news printers and publishers, resulting in some 4,300 exchange copies received by a typical newspaper each year in the 1840s. Further subsidies came from state and local governments that chose to exempt newspapers (and, later, telecommunications equipment) from their taxes. Moreover, state and local governments authorized circulation of mail with no government surveillance and supported schooling to cultivate informed citizens.

The federal government committed to ensuring that individuals and private groups had control over the content of newspapers and letters, and it deferred to states and local governance on the issue of how to educate young people. Such a commitment did not, however, prevent the government from deeply influencing, supporting, and molding the shape of the industries, technologies, and economic models for communications and media.

Government Financing and Regulation of Communications Technologies: 1840s–1940s

After the federal government financed the first telegraph line in the United States, the telegraph transmitted news of the 1844 presidential convention, and both federal and state governments provided use of lands to support a transcontinental telegraph. In later years, the federal government organized the method for distributing licenses to operate radio and television frequencies, and to this day it continues to impose only nominal fees, hence enabling private commercial interests to build business on the public airwaves.[30]

The U.S. Supreme Court stepped into action in 1936 when Louisiana governor Huey Long, smarting from criticism in the press, imposed a newspaper tax on papers with large circulations. The Court unanimously halted Governor Long's tax by concluding that imposing a higher tax on sources of public information than on other enterprises violated the constitutional freedom of the press.[31] It used the occasion to explain the purposes of the First Amendment as a commitment to ensure that information could circulate to the public, equipping the people to limit governmental mistakes and failures:

> The predominant purpose of the grant of immunity [through the First Amendment] was to preserve an untrammeled press as a vital source of public information. The newspapers, magazines and other journals of the country, it is safe to say, have shed and continue to shed more light on the public and business affairs of the nation than any other instrumentality of publicity, and, since informed public opinion is the most potent of all restraints upon misgovernment, the suppression or abridgement of the publicity afforded by a free press cannot be regarded otherwise than with grave concern.[32]

The Court rejected the tax because it was calculated "to limit the circulation of information to which the public is entitled in virtue of the constitutional guaranties."[33] Here the Court opened a view of the First Amendment as forbidding not only prior restraints on speech but also burdens on the circulation of news.

Technological innovation transformed communication during the nineteenth century and played a crucial role in supporting and expanding news and journalism. With the federal government promoting the telegraph not only through its initial investment and ownership but also through pro-business governmental policies, the telegraph became widely available, helping newspapers in the United States gain speedy and inexpensive access to news. Some may have worried that the telegraph's speed in conveying news across long distances could have undermined newspapers or reduced them to offering only opinion and commentary. But newspapers developed a network approach to breaking news and getting it "the last mile," to where the reader lives.[34]

The nonprofit Associated Press (AP), which was founded in 1846 by a handful of newspapers, now unites more than 6,000 newspapers and broadcasters in an unincorporated association.[35] It remains a cooperative service, permitted by antitrust law despite legal challenges, pooling some of the costs of newsgathering, especially in remote places, and charging other outlets for its reports.[36] Some say it also pioneered a direct and simple style for reporting news, dubbed "telegraphic." Neutral in tone and content, it was designed to be acceptable across many different news outlets. Initially spanning newspapers, and eventually radio, television, and the internet, the AP remains committed to impartial and accurate reporting. A number of competing cooperative services have developed to meet the demand for news as a product.

Although the Court ruled that collaborations could be consistent with antitrust law, at other times judicial and government enforcement of antitrust law restricted concentration of media ownership and control. Through "structural regulation," the government imposes rules restricting concentrated ownership, and it enforces the undoing of some purchases. At times the government has banned ownership of different media outlets in the same market, prohibited anticompetitive behavior by media companies, and imposed conditions on broadcast licenses and cable company authority. In each instance, governmental policies shaped the media industry, all without crossing into territory protected by the First Amendment.[37] Structural regulation affects the nature of the news gathered and reported. When the federal government allowed more concentrated ownership of newspapers during the first several decades of the twentieth century, journalism grew more professional, curating content according to emerging professional norms.[38]

From the telegraph to the internet, pro-business governmental policies have fashioned the media. Resources can come not only in dollars but in other ways, as illustrated by the federal government's grant of free use of unoccupied public lands to support development of the telegraph.[39] Government even more directly organized another new technology—radio—that would change communications forever. Private entities experimented with wireless communications in the 1890s. Radio communications had already proved significant in efforts to reduce shipping disasters, and in 1912, after fifteen hundred people went down with the ocean liner *Titanic*, the government required all ships to have wireless stations, chiefly used for point-to-point messages. Later that same year, the federal government started licensing private radio stations.

Inventors and hobbyists took the lead in using radio to communicate news. Lee DeForest broadcasted election returns in 1916 and started news broadcasts reaching a two-hundred-mile radius around New York City.[40] Within a few years, live coverage of special events, such as the Scopes Monkey Trial in 1925, showed the potential of broadcasting.

With the outbreak of World War I, the federal government sought more order in wireless communications. The government seized control of key operations in 1917, soon after the United States entered the war.[41] The government also used its purchasing power to encourage the production of radio equipment, bolstering radio's development; this approach was most actively supported by the navy, which saw the value of wireless communications across ships.[42] The government lifted patent restrictions, covered liability for patent infringements, and set uniform standards for the production of key radio components.

The business of gathering and sharing news proved resilient and even grew stronger while gradually undertaking commitments to professionalism. Competition and commercial pressures pushed publishers to expand their audiences beyond particular partisan lines, and contributed to the rise of bold graphics and sensational stories but also to investigative work.[43] Some papers even produced fake stories. In efforts to appeal to mass audiences, papers used headlines and illustrations that expanded readership. They also exposed mistreatment of the disadvantaged.[44] Even as "yellow journalism" emphasized scandals and emotional stories, journalists began to aspire to achieve objectivity in reporting.[45] When a scurrilous newspaper printed anti-Semitic claims but also exposed local government and business corruption, the Supreme Court's decision in favor of the newspaper reinforced protections for the free press.[46]

Advertising revenues replaced political party financing and motivated the search for broader readership.[47] Large urban papers published investigations of crowded housing, tainted food, and misconduct by officials, in tune with Progressive Era reform efforts. Large cities hosted multiple newspapers, many in English, and many others serving immigrants in their native languages. Economies of scale allowed lower prices per copy and higher profits; the telegraph and telephones enhanced collection and sharing of news.

Governmental policies addressed the new technologies, notably telephone and radio, and shaped their development. Using the justifications of spectrum scarcity and national security, the government built on its regulation of radio during World War I. Private investments proceeded under control and supervision by the U.S. Navy even after the war, to ensure radio's use for military defense. Over time, Congress shifted regulation from the Department of the Navy to the Department of Commerce, and then to an independent agency. In so doing, Congress maintained government direction of emerging telecommunications with a focus on serving the public interest, broadly defined.

Broadcasting has remained a largely private industry both in financing and in fundamental decisions, such as whether to permit advertising, yet a spirit of collaboration with government persisted. Privately owned newspapers invested heavily in radio but also cooperated with the state; with civic ambition, the broadcasters promised public-mindedness in exchange for light regulation.[48] With this approach, broadcasting in the United States diverged from the development of broadcasting in other countries that use governmental investment and control or a hybrid of public and private investment and control.[49]

Herbert Hoover, as secretary of commerce, set this path. He drew on his business and engineering background to foster growth of private broadcasting while orchestrating the federal licensing scheme for allocating the portion of the electromagnetic spectrum used to communicate through the airwaves. The Federal Radio Commission, started in 1927; its 1934 replacement, the Federal Communications Commission, mandated government-controlled licensing. The licenses assigned broadcasters to channels in the spectrum to avoid interference, and also required broadcasters to advance the public interest by providing local and educational content. All of these elements were subject to review by the government regulator under the authorizing language of "public interest, convenience and necessity." These words came as a suggestion from a young lawyer on loan to the Senate from the Interstate Commerce Commission, which used that exact phrase in its authorizing legislation.[50] Whatever the language was intended to mean, individual licensees and radio as a whole contributed both to mass markets and to the project of democracy.[51] By the 1940s, in order to limit market concentration in media ownership and control, the government banned mergers between local newspapers and broadcast stations (known as cross-ownership), but at other times it permitted joint operating agreements among newspapers to allow them to cut costs.[52]

By the time of the Great Depression, publishers worried that radio would draw readers and advertisers away from newspapers in the same way that newspapers and broadcasters now worry about the internet and other digital media. Broadcasting allowed journalists to give eyewitness accounts of the developments in Europe leading up to and during World War II. Radio also offered avenues for people to warn listeners of emerging fascism in Europe.[53] President Franklin Delano Roosevelt used "fireside chats" on the radio as a mechanism to talk directly to the public about his policies rather than having his words filtered through journalists. Although he did so only thirty-one times during his twelve years in office, this use of the media set a personal tone and helped overcome opposition to his plans. Here, Roosevelt pioneered use of a new technology to bypass journalists and communicate directly with people, similar to President Donald Trump's explanation for his frequent use of Twitter. During his first year in office, President Trump sent 2,568 tweets, which amounts to a little more than seven tweets each day; on one day in 2020, he sent 200 tweets. Both FDR and Trump gave boosts to the media of their times by drawing massive audiences—often boosting newspapers and magazines as well as newer technologies.

Broadcasting Licensing and Public Media

Broadcasting involved government much more than print media ever did because of the basic fact that it uses the airwaves, which are viewed as belonging to the public. Slices of the spectrum are public resources the government turned over for private use and private gain.[54] Government regulation became essential because the spectrum is finite and requires coordination. As private companies developed and marketed television in the 1950s and 1960s, dominant radio networks experimented with the new medium and built on the surge of interest in news during World War II to create news and public affairs programming.[55] An initial governmental freeze on new licenses gave leading networks (NBC and CBS) a head start. While local stations mainly relied on wire service headlines, national networks made must-see televised events out of presidential conventions, the Army-McCarthy hearings, and, later, moon launches.[56] Documentaries, interviews, and inventive formats such as Fred Friendly's televised newsmagazine *See It Now* attracted broad audiences.[57] Visuals could be captivating, instructive, and entertaining. Television generated large profits while operating within the regulatory framework well established for radio. More profitable entertainment shows could subsidize news. Over time, the immediacy of television and cable allowed those media to overtake printed newsmagazines.

Nonprofit public broadcasting, which started as educational radio using portions of the airwaves allocated by the government to noncommercial stations, emerged first with local stations in the 1950s.[58] President John F. Kennedy directed public monies to educational television and signed into law the first major federal aid to public broadcasting.[59] Congress adopted the All Channels Act in 1962, requiring manufacturers of television sets to include VHS tuners, which increased the number of broadcasting channels and opened room for public educational broadcasting channels in the many communities without one.[60] National public television launched in 1967 with a nonprofit organization serving as a buffer between local stations and the federal government. National Public Radio started in 1970, broadcasting Senate hearings on the Vietnam War and developing independent news reporting, funded partially by the government but mainly by private donations. Federal support for public radio, television, and web-based media, while always a modest part of the government's annual budget, provides a basis for matching private donations with public funds, and government support is amplified by the tax-deductible treatment of private donations.[61] Government thus framed the development of public media and contributed resources directly and indirectly while encouraging private support.

Satellite, Cable, and the Internet

Communications options increased with the ability of communications satellites—authorized by Congress in 1962 to permit the commercialization of space-based communications—to relay signals across widely separated areas. The technology supported and transformed telephone, radio, and television communications. Private companies also invented cable technology, which carried television signals to remote places.

Conflicts between broadcasters and cable operators drew in the courts. Creating some order in the context of competition, the FCC extended its regulatory reach to cover cable. Initially the FCC required cable systems to carry local stations, and later it prohibited during specific and limited time frames the importation of programs from nonlocal stations where those programs duplicated those on local stations.[62] Over time the FCC enforced fewer regulations on cable, while state and local governments established franchising authorities to govern the industry.[63]

The rapid expansion of satellite distribution and cable services threatened the networks, but network news adjusted, and cable found audiences seeking news 24/7. Ted Turner's CNN and Rupert Murdoch's Fox News demonstrated that there were sufficient audiences for continuous news shows, though Fox brought more edgy, populist, and entertaining elements, drawing on popular talk radio as an alternative to what some viewed as a liberal bias at CNN. MSNBC further fractured audiences with a more explicit liberal slant.[64] And by 1999, Comedy Central drew in younger viewers with Jon Stewart's comedy program *The Daily Show*, which offered a sharp satiric take on the news, especially politics. The development of news aimed at conservatives and news aimed at youth interested in comedy reflected the viability of the cable industry as it grew under the oversight of federal, state, and local governments.

Government Funding of Research and Development

Government-directed and government-funded research underlies critical innovations in communications and computer technologies, just as was true with transportation networks in an earlier era.[65] Federal dollars covered at least 50 percent of research and development investment in the United States between 1951 and 1978, the crucial period in developing telecommunications and the internet.[66] Government investment embodies not only material resources but also great patience and staying power through the long periods of experimentation and failure involved in the innovation process. Private enterprises bring investment and risk-taking, but often they harvest the results of significant government vision, direction, and financing of breakthroughs. A prime example: Google's basic algorithm was developed with a National Science Foundation grant.

In fact, government involvement in the development and design of the internet has been direct and significant, with profound consequences for the ecosystem of news. It was the federal government that supported the creation and preservation of an open and accessible internet. The federal government subsidized research on and demonstrations of computer networking that became ARPANET, the precursor of the internet.[67] Over time, the federal, and especially state and local governments accord significant subsidies to dominant internet platforms.[68] The federal government also guaranteed consumers the right to use modems on their phone lines and prevented telephone companies from undermining the emerging computer network market. After Congress overhauled telecommunications law in 1996, the federal government required phone companies to share their lines with competing broadband services (DSL) enabling internet access through phone lines.[69] Congress has worked to promote speech on the internet—but also, with judicial interpretation, set some guardrails.[70] Governmental decisions to forbid or to permit internet service providers (such as Comcast and Verizon) to speed up services for some and slow down or even block service for others shape the availability of news and other information. Debates over whether government should allow such discrimination or instead require "net neutrality" have occupied both federal and state regulators.[71] Meanwhile, Section 230 of the Communications Decency Act provides a form of subsidy for platform companies by insulating them from the kinds of liability for falsehoods, defamation, and other legal violations that publishers and legacy media deal with every day.[72]

The point of this historical narrative is not simply to show how private companies as well as ultimate consumers benefit from research backed by the taxpayers' contributions and patents authorized by Congress. Government instigation, resources, oversight, and influence have been *indispensable* to the development of modern communications. While private enterprise has supplied financial and managerial resources and deeply influenced media and news, federal antitrust policies and practices and federal communications regulation have played powerful and essential roles in the shaping of the nation's news ecosystem.

The nation's news ecosystem now includes interactive forms of content collection, production, and distribution. Networks and cable channels developed web-based journalism, offering in-depth features, visuals, and other follow-up material that strengthened television news reporting. Sophisticated analyses identify how to engage audiences and push out news using social media and other platforms, as well as through traditional media.[73]

Over time, this pattern of technological innovation has upended old industries but also opened new ones. Radio, film, and television affected the sources and forms of live theater but did not eliminate its audiences.[74] "Streaming may have killed home video," observes Matt Pressberg, "but it ended up ushering in a different kind of boom of watching movies and TV shows at home."[75] Digital distribution of music first led to piracy and plummeting sales, but then iTunes, Spotify, and Pandora developed ways to charge for their product and expand audiences; it took twenty years, but now the music industry is increasing revenue.[76] Old media may disappear, but it also may be reinvented with new purposes alongside innovations, just as radio persisted after the rise of television and the internet.[77] Alongside opportunities for entertainment, communications technologies opened avenues for investigative journalism, documentaries, and inventive ways to combine news with humor and art. And while early internet pioneers may have dreamed of providing endless opportunities, the new media platforms have often wound up imposing new constraints.[78] In the age of digital communications, the central scarcity is the attention of listeners, who face floods of messages and communications and often lack the sufficient tools to make their own choices about what deserves their time.[79] Newspapers, broadcasters, and cable companies have struggled to adapt to the digital age as internet companies pursue markets without particular interest in journalism and its ethical and professional aspirations.[80]

TECHNOLOGICAL CHANGE AND THE PRESS

Throughout periods of technological change that have profoundly increased communications possibilities, the U.S. government presumed that a private press would exist, while often assisting it by regulating the relevant infrastructures and private enterprises. Federal, state, and local governments in the United States have never been hands-off when it comes to the media. Early postal subsidies for newspapers; investments in telegraph, radio, television, and internet technologies; and regulation of airwaves, satellites, and cable are only the most visible forms of government involvement—there are others. Take copyright law enforcement: it stymies some new communications options while supporting others.[81] And the government has long subsidized commercial broadcasting by levying only a nominal fee for radio and broadcast licenses and by treating corporate advertising as one of a range of deductible business expenses. Tax treatments, broadcast licensing, and antitrust regulation coexist and shape private media and a conception of news and other media as independent of government. Government efforts directly and indirectly for more than two hundred years have supported that conception.[82]

The federal government and telecommunications businesses have often produced uneasy partnership. Not infrequently, the government has threatened to hit telecommunications companies with more regulation. Conversely, critics can point to industry influence on the government or even argue that private interests have at times "captured" governmental regulation. That claim only underscores how the direction and development of media reflect ongoing government involvement.

Through antitrust policies and enforcement practices, the U.S. government at times promoted competition in services and ownership among news media and all telecommunications but at other times shielded private investors in the United States from international competitors.[83] The big decision to take apart the largest telephone company (AT&T) through antitrust litigation starting in 1974 and settling in 1982, marked a government judgment to prefer innovation through competition over the quality and reach of services ensured by a regulated monopoly.[84] Federal policies have also promoted universal service and steered private industries toward that goal.[85] At other times, federal antitrust policies permitted concentration of ownership and consolidation in media. The rules and enforcement actions governing media competition shaping the types of choices available to consumers and addressed market failures.[86] Through the Newspaper Preservation Act of 1970, Congress allows newspapers in the same market to coordinate on business functions such as printing and advertising as long as they keep their editorial functions separate.[87] Whether or not it is wise policy, this law shows the government's willingness to alter antitrust policies in order to strengthen media diversity.[88]

Over time, "the press" protected by the First Amendment has come to encompass print media, broadcast and cable media, and now the internet. Rules and practices issued or influenced by government shape interstate communications of radio, television, wire, satellite, and cable. Government efforts to regulate the internet in the United States and elsewhere have remained contested. Growing arguments for regulation of internet-based companies generate tension between internet platforms and the United States as well as other governments. Yet the platforms rely on governments to enforce their rights and the government at times relies on the companies in carrying out its law enforcement and cybersecurity functions.[89] Media, including news gathering and distribution, largely flourished in the United States with constant involvement of legislation and administrative policies as well as judicially enforced constitutional protections of private property and freedom of speech and government promotion of competitive economic and technological development. The constitutional plan did not only assume the existence and viability of private enterprises producing and distributing news; it also authorized governmental contributions to the news industry through decades of economic and technological change. It is a history of many disruptions and shifts. The news business in particular changed with the telegraph and telephone, radio and television, cable and wireless, the internet, computers, and mobile phones. These technological changes and the transformations they brought did not occur in a vacuum. Over the course of U.S. history, media grew with subsidies from government actors (including the military) and political parties, with advertising and direct-to-consumer purchases, and with an increasingly widespread view among the public that the media should be a pluralistic, independent watchdog of the state.[90]

Over American history, media businesses competed, and at times they cooperated—all subject to government concerns about economic concentration and consumer exploitation. News industries have depended upon a mix of private financing, including advertising, with consumer subscriptions and government subsidies for both private enterprises and nonprofit ones. The federal government has subsidized, regulated, and shaped media technologies with policies helping media prosper without domination by a few players and, for the most part, without government censorship.

Are the disruptions affecting news since 2000 more severe or different in kind than the prior disruptions of telegraph, radio, television, and cable? Serious financial problems exist for legacy media, especially newspapers, which face debt, labor and distribution costs, and constant competition from newer media and internet entities. The central problem now is not governmental overregulation curtailing freedom of speech but inadequate government involvement to prevent domination by a few companies and the swamping of users with a plethora of messages, propaganda, memes, and ads. At stake is not just how news media can find paths to financial viability but how to maintain the free expression of news and opinion, called by the Virginia Declaration of Rights in 1776 "one of the greatest bulwarks of liberty."[91] Yet a libertarian conception of the First Amendment, which has become more prominent during the past several decades, is at odds with the historical role of the government in the shaping of the news industry—and puts the prospects of government action at risk.

CHAPTER THREE

. . .

Does the First Amendment Forbid, Permit, or Require Government Support of News Industries?

DOES THE FIRST Amendment forbid reforms to save the viability of newsgathering, production, and distribution, and to guard against the forces undermining both the news industry and the access to information that ordinary people need to self-govern? Does the Constitution prevent governments from restricting internet platforms that fail to pay for news gathered and produced through newspapers and other media and also fail to guard against deliberate misinformation or hateful expression? Can exposure to a variety of views be a lawful requirement for algorithmic news feeds? Might some efforts to improve the ecosystem of news be not only permitted but propelled by the First Amendment? As is the case with so many important legal issues, the most likely answer is, "It depends."

The assumption that the First Amendment bars government from playing a role in media systems and the markets for news—called "First Amendment fundamentalism" by media studies scholar Victor Pickard—hobbles public debate.[1] Judicial interpretations over time and legislative and regulatory enactments permitted by the courts demonstrate ongoing government involvement with media and news industries rather than some simplistic rule against making any law affecting expression. Judicial interpretations have found a variety of government interventions compatible with the First Amendment and defy any assertion that the government is constitutionally required to stay away from the news and media worlds. No doubt, actual development of First Amendment law turns on the specific shape and device of any possible government action. Also, recent decisions point toward more aggressive interpretations of the First Amendment as a tool against corporate regulation, but the scope of this trend has yet to be decided.

In light of the First Amendment, government actors face significant restrictions, especially around the choice of expressive content, and yet the federal government also significantly shapes media and hence expression. The Constitution need not foreclose government action to regulate concentrated economic power or to support news initiatives where there are market failures as long as the regulation stays clear of limiting who can speak or select the speech. Thus, while the Supreme Court has rejected restrictions on campaign expenditures by corporations and nonprofit organizations, it has permitted enforcement of laws against monopoly or oligopoly economic power.[2] If past Supreme Court interpretations of the First Amendment persist, government can continue to require disclosure about who is financing particular communications. Given present developments, a strong case can be made that public policy can protect users from bombardment by computer-generated messages and implement other reforms designed to screen out uninvited distractions.[3] Moreover, the First Amendment may itself require the government to take action to strengthen the information and news functions presupposed by democratic governance. This is admittedly a bold claim beyond existing judicial decisions, but it is in line with other constitutional claims that some affirmative steps by government—such as educating children—can be required in order to ensure the preconditions for constitutional democratic governance.[4]

Law professor Tim Wu argues that the First Amendment, which was intended to prevent government censorship, is becoming irrelevant to this sprawling world of private companies and deceitful agents.[5] He argues that responsibility falls to private companies, technologists, and legislators. Indeed, all of those players should step up, but the First Amendment and the Constitution generally remain implicated, relevant, and motivating. The First Amendment constrains Congress from *abridging* the freedom of the press and the freedom of speech, but it does not bar actions to *strengthen* them. To sustain freedom of the press and enable people to participate in democratic governance, courts need to pursue some new approaches, and so do Congress, the states, and private actors. At the same time, the global nature of contemporary information gathering and distribution means that more than one nation influences the rules governing digital platforms. As a result, potentially conflicting legal regimes and values are at work. In the near future, constructions of the First Amendment might well take into account what is needed to preserve a meaningful role for the United States in the regulation of U.S.-based internet platforms.

The language and traditions of the First Amendment rightly stress the dangers of government control of expression. The risk of government officials censoring opinions or investigations that expose the corruption or failures of the government requires constant vigilance. That vigilance does and should extend even to affirmative governmental assistance to communications. Yet past and potential constitutional treatments show the compatibility of the First Amendment with common-law and statutory protections against defamation and fraud; with reforms affecting the economic structure of industries and their technological predicates; with taxation and subsidies; with consumer protection, intellectual property, and other liability rules; and with government-led efforts to improve the experiences of users. Users (once known as readers, listeners, and viewers) now can easily contribute as well as receive news content—and, knowingly or unknowingly, they contribute their personal data, which are used to finance digital enterprises. Justifiable concerns about government constriction of speech should not impede government protections of individuals as consumers, as receivers of information, and as participants in self-government.

Key to understanding the First Amendment's meaning are the decisions by courts and legislatures that shape the industries and the mechanisms for gathering, producing, and distributing news. The place to begin is with a sketch of the historical shifts in First Amendment interpretations.

FIRST AMENDMENT DEVELOPMENTS: FROM NON-ENFORCEMENT TO ROBUST PROTECTION

Shifting meanings characterize the First Amendment over time, a phenomenon that makes attention to the amendment's purposes and to new contexts always critical. For the first century of the nation, the freedom of news and opinion enacted in the First Amendment's protection for speech and for the press expressed values but gave rise to little legal articulation, including by judicial enforcement.[6] Nonetheless, the Constitution's framers and early national leaders discussed the ideals represented by the First Amendment. After his service as president, Thomas Jefferson reasoned that "where the press is free, and every man able to read, all is safe."[7]

The First Amendment expressly directs its restrictions only to Congress ("Congress shall make no law . . . abridging the freedom of speech, or of the press"). As a result, state and local governments could seem beyond its reach.[8] When first presented with the question, the Supreme Court rejected arguments for applying rights announced in the federal Bill of Rights to actions by state governments.[9] This did not seem surprising. The Constitution's framers appear to have worried more about ensuring state power than about protecting discourse from government strictures.[10] Common-law rules about defamation and libel, for example, remain enforceable even after the adoption and expanding judicial interpretation of the First Amendment. Despite criminal libel laws on the books, however, the press proceeded in the early republic to provide vigorous criticisms of government.[11]

False and negative comments still give rise to legal liability, but there is greater latitude for comments about government officials and other persons pervasively involved in public affairs.[12] Truth as a defense to common-law charges of libel or defamation emerged through the federal Sedition Act of 1798. Such a defense may have had little practical value for writers and speakers using epithets and expressing opinions, yet even so, with this legislation the United States ensured greater protection for speech than the scope traditional in England.[13] That same Sedition Act, enacted by the Federalists against their Democratic-Republican opponents, criminalized more than current understandings would allow by providing for penalties against individuals acting in conspiracies "to oppose any measure or measures of the government" or making "any false, scandalous and malicious writing" against Congress or the president.[14] As a candidate for president, Democratic-Republic Thomas Jefferson denounced the laws and he secured election in 1800.

Besides political responses to curbs on speech and the press, state constitutions and other state laws did and still do offer protections. So the practices stood until after the Civil War and the enactment of the Fourteenth Amendment, whose clause protecting life, liberty, and property from deprivations by the state gave rise to judicial interpretations construing some (but not all) of the federal Bill of Rights as applying to the states.[15] This process of "selective incorporation" of the Bill of Rights to bind actions by state governments meant that legally enforceable protections of speech and the press became available in federal court but in practice depended upon rules within individual states until the twentieth century.

Between 1917 and the 1930s, active campaigns by fledgling civil liberties initiatives pushed back against wartime and other restrictions on speech and on the press, and protections of speech even during wartime strengthened.[16] Skeptical of the courts, which had issued injunctions against labor picketing and strikes as violations of conspiracy and vagrancy laws, labor leaders gradually began to support litigation efforts pursuing judicial protection for freedom of expression.[17] When labor and fledgling civil liberties groups pressed challenges to government suppression of speech, a minority of justices on the Supreme Court found the arguments compelling. Dissents by Justices Oliver Wendell Holmes Jr. and Louis Brandeis pushed against punishments for antiwar publications and other forms of dissenting speech and laid the ground for majority rulings that eventually expanded legal protections for expression.[18]

The justices leading these changes took different paths. Justice Holmes actually changed his mind about the scope of the First Amendment; he departed from his usual defense of legislative majority rule and objected to punishing political activists for their speech. Perhaps he was influenced by seeing some of his friends caught up in repression; correspondence with thoughtful defenders of robust protectors of speech no doubt contributed to his change of heart.[19] Justice Brandeis consistently defended enforcement of the First Amendment against the states. As a lawyer, he had relied on the press to expose government lying and private misconduct, and he perceived the necessity of freedom of expression and assembly for labor organizing and for the education and spread of the information people need for self-government.[20] Having played an important role as an advocate for unpopular causes, Justice Brandeis conceived of the First Amendment as essential to a society founded in and supporting political deliberation, a society depending upon (and helping to develop) free individuals able to make good decisions.[21] The dissenting opinions by Justices Brandeis and Holmes framed the doctrine ultimately embraced by the Court's majority: the federal courts would have a role in evaluating and striking down state legislation and state law enforcement. In so doing, the federal courts became terrains for disagreements over speech restrictions and the scope of judicial power concerning avenues of communication and the availability of news and information.[22]

The real shape of constitutional protections comes through ongoing debate across society in different historical periods. Wartime expanded government powers to restrain speech, with judicial approval.[23] Educational institutions, the media, the legal profession, and civil liberties organizations advocated for broader interpretations of the First Amendment and crafted stories about the lessons of history regarding freedom of expression. Despite arguments for the special rights of the press, the Supreme Court has resisted claims of the need to keep journalists' sources confidential or to grant the press special access to closed government spaces. Instead, the courts folded the press into the protection for speech by anyone, and over time they broadened general judicial protection for freedom of speech.[24] Over time, judicial decisions expanded protections of speech to include honest mistakes in comments about public officials.[25]

New technologies prompted new interpretations. Courts ruled that First Amendment rights extend to broadcasters, but the strength of the First Amendment's shield against regulation is less robust for them than for some other forms of speech.[26] The scarce resource of airwaves and the requirement that government licenses be obtained to use them justified greater restrictions on broadcasting than on print. Despite assertions that anyone who can purchase airtime should be able to do so, broadcasters who secure government licenses are constitutionally accorded power to select whose voice gets on the air. Editorial control over programming is part of the First Amendment freedom and licensees' rights—though it is subject to federal regulation seeking to require that uses of the airwaves be consistent with the "public interest, convenience, and necessity."[27] Charged with protecting the public interest, the licensees remain subject to government oversight. Broadcasting practices developed within a context of regulatory constraints, competition, and professional practices that prized high editing and production standards while serving—or claiming to serve—an idea of democratic participation and national connectedness.[28]

Federal regulation of broadcasters has existed with constitutional approval. The Supreme Court allowed government regulation, and specifically permitted regulation of broadcasters to include the Fairness Doctrine formally announced by the FCC in 1949. Rooted in a 1929 denial of a license to a radio station controlled by a labor union, the Fairness Doctrine reflected the view that broadcasters hold a public trust to ensure that competing views receive ample airtime. Broadcast licensees, entrusted with the scarce resource of the spectrum, can be required by the government to present competing sides of controversial issues of public importance covered in their broadcasts, in order to serve the public interest. The scarcity of broadcasting licenses—matched to a limited number of electromagnetic frequencies—justified governmentally imposed duties on license holders to present a variety of views and also a right of reply for those attacked on the air.[29] For quite some time, media industry leaders supported these ideas as enforced in the Fairness Doctrine.

The Supreme Court has agreed. In its 1969 decision in *Red Lion Co. v. Federal Communications Commission*, the Supreme Court upheld the Fairness Doctrine despite a constitutional challenge.[30] The Court pointed to the governmental role in allocating broadcast frequencies and to the legitimate claims of possible users. The Court also upheld "must-carry" requirements for news, cultural, and current affairs programming as advancing legitimate aims rather than suppressing content. The government's oversight of the selection process for broadcast licenses expresses a public concern with the structures affecting media and departs from the alternative of a purely private property approach. Local governments' choices of cable companies authorized to operate in their jurisdiction—and the delegation of power over this choice to the local governments in the first place—reflect public policies that influence competition, market power of companies, content of media, and choices of consumers. The Constitution does not bar government involvement here. For example, a statutory requirement of reasonable access to free or paid time for political candidates does not violate the First Amendment and instead reflects the responsibilities of broadcast licensees.[31]

Although it is no longer in force, the Fairness Doctrine deserves renewed attention. Its judicial treatment shows that the First Amendment can be compatible with government promotion of quality media content, including national and local news, the arts, and educational content, through the enforcement of duties placed on broadcast license holders.[32] Even though the government largely justified of the Fairness Doctrine and the mandated right of reply on the scarcity of broadcast frequencies used by television and radio licensees as trustees for the public, some of the reasoning remains relevant to other contexts.[33] Thus, in *Red Lion*, the Supreme Court upheld the equal-time provisions of the federal government's Fairness Doctrine in light of "the right of the public to receive suitable access to social, political, aesthetic, moral, and other ideas and experiences."[34] This idea echoed earlier commitments in the United States to ensuring an informed electorate, well documented by historians.[35] The justices unanimously upheld the Fairness Doctrine by interpreting the First Amendment as protection for "the right of the viewers and listeners, not the right of the broadcasters." This helped justify the government requirements that broadcasters provide "ample play for the free and fair competition of opposing views."[36] Even outside of this context, it makes sense to understand this public right to receive information as itself integral to the First Amendment.[37] The Supreme Court's recognition of the significance of cyberspace and social media to communications includes great skepticism over governmental exclusions of anyone—even a convicted sex offender—from freedom to use those tools.[38] The right to receive information underscores the fundamental role of the First Amendment in a democracy that requires active and informed participants.

In some ways, the Fairness Doctrine echoed an ideal of journalism first advanced by founding father Benjamin Franklin. He defended the service afforded to the public by the printer-editors who would be expected to open their pages to contrasting points of view, much as a ship captain or coach operator would transport individuals of different persuasions.[39] A similar spirit animated the Fairness Doctrine, permitting broadcasters to express particular positions but requiring them also to provide a platform for others to do so, with the goal of reasonably balanced expressions of competing views.

At one time, the Federal Communications Commission officially described the Fairness Doctrine as the "single most important requirement of operation in the public interest—the sine qua non for grant of a renewal of license."[40] The Supreme Court approval of the Fairness Doctrine amplified an ideal of nonpartisanship and neutrality embraced by some leading broadcasters even though balance at times is elusive or inapt. False equivalences between histories of the Holocaust and Holocaust deniers and between climate scientists and those who say climate change is a hoax should not be justified in the name of "balance." Edward R. Murrow observed in the 1950s when Senator Joseph McCarthy targeted named government officials and others as Communists: "Some issues aren't balanced."[41] As Senator McCarthy prepared to attack Murrow for his reports, Murrow devoted an episode of his show to a sustained examination and attack on McCarthy and in defense of free speech and dissent. Criticism of government officials of course lies at the core of the First Amendment, and lack of balance is easy to see when a government official dominates the news without any expression of contrasting views.

Political support for the Fairness Doctrine weakened amid attacks by the industry. The government terminated the policy in 1987 as President Ronald Reagan's deregulation policies took hold. The Federal Communications Commission removed the Fairness Doctrine rule from the books along with others deemed no longer necessary.[42] Broadcasters no longer had to provide a balance of comments on an issue. The change in part reflected deregulation and a pro-business ideology. Support for the rule persisted, though. Congress tried to restore the rule, but President Reagan vetoed it. The rise first of cable and then of the internet altered the regulatory predicate of scarce speech opportunities and to some, reduced the need for a policy requiring balance within one outlet. Yet a deeper explanation for the end of the Fairness Doctrine lies in the erosion of public interest ideal in media and in the country as a whole.

The political end of the Fairness Doctrine opened avenues for radio and television to seek to advance a particular point of view aimed at a slice of the community.[43] Talk radio became a particularly favored medium for right-wing pundits. Rush Limbaugh, whose nationally syndicated show launched in 1988, offered a point of view (a conservative one) on events rather than original reporting. The success of right-wing broadcasting paved the way for Fox News, as Rupert Murdoch in 1996 empowered Republican media consultant Roger Ailes to pursue media with no ambition of serving all people; instead, the plan was to advance a particular viewpoint.[44] Framed in part as an alternative to CNN, the first 24/7 news outlet, Fox News recruited prominent conservatives and maintained strong ties with the Republican Party. CNN, which had been launched in 1980, exemplified a commitment to broadcasting competing views, with shows such as *Crossfire* and *Evans and Novak*. Those shows, drawing many viewers, presented debates over topical issues.

Fox News had a different, deliberately partisan approach.[45] Gaining popularity over time, Fox became the most trusted source for some and the least trusted for others—exemplifying and fueling the polarization of both media and politics in the United States.[46] In retrospect, both debates over and then the demise of the Fairness Doctrine seem to have contributed to the more divided media environment of the first decades of the twenty-first century, whether or not that was anyone's intention. Although a real understanding of how such division comes about will require more empirical study, polarization may increase because of the ways dominant digital platforms structure news feeds and social media.[47] Ideologically driven media contribute to the lenses people bring to whatever they learn, leading different people to draw quite different conclusions from the same material.[48] Even though the viewership of cable news remains smaller than for broadcast news and social media accessed by mobile devices, its divisiveness draws attention.

The end of the Fairness Doctrine contributed to the current situation. It was a political demise, not a legal one. When applied to broadcast media (as opposed to print media), the Fairness Doctrine and its constitutionality had not been overturned by any court. A fair question is whether it would remain viable legally as the predicate of spectrum scarcity fades, given that content is now carried not just by broadcasting but also over cable and the internet, carried across Wi-Fi, fiber optics, broadband, and other means. It is also fair to ask whether a communications world consisting entirely of competing partisan, ideological speech is worse than one with at least some sources aspiring to distinguish fact from opinion and to bring competing views into direct exchange. The partisan press of the early nineteenth century grew into a mature industry with professional standards of fact-checking and distinguishing editorial and reporting work. As consumers and users criticism current media for bias, the disruption of responsible journalism reflects economic and technological disruptions rather than a disintegration of its ideals.

While government should not intrude on the private reasons of editors and speakers who choose what to express, regulation of media carried through technologies other than broadcasting has been and can be permitted under the First Amendment. Government can attend to anticompetitive practices and can regulate deception and abuse of consumers' trust.

Even without the rationale of spectrum and license scarcity, decency has been the focus of lawful regulation. The Supreme Court has reasoned that the federal government can authorize a cable operator to enforce prospectively "a written and published policy of prohibiting programming that the cable operator reasonably believes describes or depicts sexual or excretory activities or organs in a patently offensive manner as measured by contemporary community standards."[49] A congressional effort to protect children from sexually explicit speech on the internet met with three rounds of litigation and two Supreme Court decisions, which ultimately left in place an appellate court's halt of the law.[50] In light of judicial skepticism about content-based restrictions, the government changed regulations of cable or internet communications to content-neutral (and viewpoint-neutral) approaches or else rejected the restrictions altogether.[51] Nonetheless, the FCC still limits obscenity, indecency, and profanity during certain broadcasting hours.[52]

And the Supreme Court has upheld a must-carry provision requiring cable networks to carry public educational stations. Applying intermediate scrutiny, the Court held that the law served governmental interests, citing the interests in competition, information dissemination, and "preserving a multiplicity of broadcasters."[53] Also long-standing, and still in effect, is required disclosure by both broadcasters and cable operators of the actual sponsors of materials, because "listeners are entitled to know by whom they are being persuaded."[54] As an effort to avoid spreading harmful hoaxes, the federal government established that broadcasters also must not present information that they know to be false, that they foresee would cause serious harm, and that does in fact cause serious harm.[55] While members of the Supreme Court have acknowledged that some falsities need protection in order to ensure a flourishing marketplace of ideas, they have also emphasized the lack of value in false factual statements that can interfere with the truth-seeking function of this marketplace.[56] These harms are exacerbated in digital communications, where misinformation campaigns face few costs and editorial activity is turned over largely to algorithms designed to elevate human engagement.

By the early part of the twenty-first century, the federal government had authorized light government oversight of technology in the hands of private businesses in addressing speech issues and public concerns.[57] Governmental efforts to promote quality media—including national news, local news, educational content for children, and the arts—have a long history in the United States and have not been found to violate the First Amendment, despite warnings by some critics that any government connection jeopardizes freedom of speech.[58]

The First Amendment has provided courts, legislators, and other actors a terrain for argument and analysis, with striking shifts during different eras resulting in changes both in what news is conveyed through varied communications domains and how it is conveyed. Evolving from deference to state authorities, current First Amendment doctrine more aggressively restricts government regulation of speech. Yet the Supreme Court has also often acknowledged not just the rights of speakers but also the rights of listeners, and government actions may be warranted to ensure that listeners have choices and knowledge about those choices.[59] All these constitutional elements persist at a time when individuals have never more easily or cheaply communicated so broadly—and when professional journalists are rapidly losing jobs, facing threats to livelihood. The curious juxtaposition of the dying industry of newspapers and expanded judicial protection of freedom of speech—embracing cable television companies, advertising, corporate campaign contributions, trademarks, and computer code—should provoke renewed attention to the First Amendment's place in the constitutional scheme. Where are the rights of listeners as well as those of speakers? Where is attention to the overall structure of communications and the gathering and distribution of news required to enable an informed electorate and to hold governments accountable? At a time when the actual business of the press on the ropes in communities across the nation, the courts have adopted an increasingly libertarian approach to the First Amendment. The motive is not, however, addressing the role of news media in a democracy but instead pro-business deregulation pushed by business interests unrelated to journalism.

EMERGENCE OF THE LIBERTARIAN—AND "WEAPONIZED"—FIRST AMENDMENT

Corporate and anti-union advocacy groups began to use the First Amendment to challenge any regulation. They have started to see success, beginning in 2005 in the Supreme Court led by Justice John Roberts. A 2018 Supreme Court decision ruled that government employees who are represented by a labor union but do not belong to that union cannot be required to pay a fee to cover the union's costs to negotiate a contract governing all employees. In *Janus v. AFSCME*, a majority on the Supreme Court announced that government employees have a fundamental First Amendment interest in choosing whether to support a union and the public sector labor negotiation system that determines their own wages—and thus used the First Amendment to strike down as unconstitutional a key part of the framework for labor-management negotiations. In so doing, the Supreme Court overturned a precedent of forty years.[60] The Court also deployed what can be called a "libertarian" First Amendment interpretation to undermine economic regulation. The decision to enlarge First Amendment protections this way is especially perplexing because in recent years the Court has shown little interest in the freedoms of employees to speak their own views; for example, the Court refused to recognized the free speech rights of a state prosecutor who distributed a questionnaire about office transfer policy and pressures to work in political campaigns.[61] In decisions expanding protections for corporate speech and rejecting campaign finance reforms on the view that money in campaigns is speech, a majority on the Supreme Court has accepted and extended anti-regulatory uses of the First Amendment.[62]

In dissent in the *Janus* case, Justice Elena Kagan argued that the majority "weaponized the First Amendment, in a way that unleashes judges, now and in the future, to intervene in economic and regulatory policy."[63] Also dissenting, Justice Stephen Breyer analogized this kind of construction of the First Amendment to the aggressive Supreme Court use of due process more than a hundred years ago to strike down economic regulations.[64] Ushering in one of the most controversial—and later rejected—lines of constitutional analysis, a majority of justices ruled in *Lochner v. New York* (1905) that limits on workers' hours violated the Fourteenth Amendment's protection of individual liberty and property through the guarantee of due process of law.[65] The case was decided by one vote. Among the dissenters was Justice Oliver Wendell Holmes Jr., who stressed that the Constitution does not enact laissez-faire or any other particular economic theory. The dissenting view ultimately prevailed when the Court upheld New Deal labor protections put in place after the Great Depression.[66]

Once again, a divided Supreme Court seems to be using the Constitution to put into practice a particular theory skeptical of government regulation, but now the vehicle is the First Amendment.[67] In this vein, in *National Institute of Family and Advocates v. Becerra*, the Supreme Court ruled by a vote of 5–4 in favor of a First Amendment challenge to a California law requiring crisis pregnancy centers—viewed by critics as efforts to use deception and misinformation while pretending to provide a wide range of services—to provide visible notices of the availability of abortion and other options at state-sponsored clinics.[68] Antecedents exist for the aggressive use of the First Amendment to reject neighborhood, employee, and consumer protections; during the 1930s–1940s and again in the 1970s, the courts at times interpreted the First Amendment to reject local taxes on door-to-door peddling.[69] But this minority view of the First Amendment has long been identified with the failed and rejected approach of the *Lochner* Court, substituting the judgment of jurists for democratically enacted economic regulations.[70] It also is a minority view that restricts or ignores the First Amendment rights of listeners and viewers to receive information.[71]

Explaining these developments, legal historian Laura Weinrib notes how in the midst of economic stagnation in the 1970s, advertisers, political donors, and eventually employers chipped away at the reach of the regulatory state."[72] Jeremy Kessler, another law professor, argues that activists now are pushing a libertarian suspicion of economic regulations.[73] Shifting membership on the courts may yield more openness to these arguments, curbing or preventing government action, despite precedents going back half a century according government many tools to strengthen or save the gathering, production, and distribution of news. If emerging judicial interpretations of the First Amendment prevent the government from addressing the erosion and collapse of newsgathering, reporting, and distribution, those potential judicial interpretations deserve serious critique. The Constitution is "not a suicide pact."[74] Especially if emerging judicial constructions of the First Amendment cast doubt on government subsidies for public media, enforcement of antitrust, consumer protection, intellectual property laws, and requirements of fair and nondeceptive curation of content, such judicial interpretations would run afoul of long-standing precedents as well as of the Constitution's presumption of a viable press supporting an informed citizenry.

Some see the capture of the First Amendment by businesses and trade groups ever since the mid-1970s, when the Supreme Court announced protection of "commercial speech."[75] Corporate and business organizations now regularly and often successfully challenge regulations on consumer, business, and securities trading practices.[76] Professor John Coates conducted an empirical examination and concluded that "nearly half of First Amendment legal challenges now benefit business corporations and trade groups, rather than other kinds of organizations or individuals."[77] Coates also analyzed and debunked claims by the courts that First Amendment protection helped create the period of great economic growth in the United States between 1850 and 1960.

Nonetheless, even governmentally required factual disclosures to serve the interests of consumers and citizens are at risk now. Under pressure is a 1985 Supreme Court decision that a state can require commercial speech to include "purely factual and uncontroversial information" without violating the First Amendment "as long as the [state's] disclosure requirements are reasonably related to the State's interest in preventing deception of consumers."[78] For now, the requirement that federal political candidates disclose their expenditures remains consistent with the First Amendment, but required disclosures regarding political ads face challenges.[79] Yet existing law requiring disclosures—such as disclosure of the fictional nature of what might otherwise be viewed as a dangerous hoax—could face constitutional jeopardy.[80] Requiring an online site to disclose who paid for a political ad violates the First Amendment, according to one court, whether or not the standard of review is stringent.[81] The status of other required disclosures is unclear because of shifting ideas about what counts as unconstitutionally compelled speech—forbidden governmental actions that force an individual to express or support certain speech.[82]

Regulators outside the United States have begun to mandate disclosures by U.S. tech companies. If the United States follows suit, expect challenges in U.S. courts.[83] This augurs trouble for initiatives requiring platform companies to disclose information about their sources, algorithmic tools, uses of consumer data, or digital interfacing information, but nothing would halt voluntary disclosure by such companies—even if those voluntary disclosures are encouraged by a variety of other policy tools. Addressing these and other avenues open for public action requires a closer constitutional analysis of governmental tools. The text of the First Amendment does not change, but political power does, and interpretations by judges appointed through a political process can shift in turn. Courts attend to the differences among media and historical contexts. Division among federal courts over the very standards of review to use in the context of government-mandated disclosures of the sources and sponsors of their content creates opportunities for more nuanced and thoughtful approaches, such as proportional balancing of harms with the underlying purposes of the Constitution.[84] The parameters of specific past judicial decisions offer some guide to avenues for action in the present; they also suggest areas that could and should change.

CONSTITUTIONALITY OF GOVERNMENT TOOLS WITH POTENTIAL TO STRENGTHEN OR SAVE NEWS INDUSTRIES

Since the origin of the United States, federal, state, and local governments have participated in decisions shaping and subsidizing the channels of communication through which news is gathered, produced, and distributed. Those actions have long taken place with minimal restrictions traceable to the First Amendment. Government assistance to the media has long been deemed compatible with or even required by the Constitution: the First Amendment has been read to permit subsidies to media and also to protect the public's right to access information needed to participate in democratic self-governance and other activities. Judicial decisions preserve the latitude for the government to guide and support a robust news and media ecosystem. The First Amendment coexists with antitrust law, tax law, government subsidies, disclosure requirements, and consumer protection, as well as with intellectual property law and libel and defamation laws.

Intellectual Property and Defamation Laws

Congress has immunized interactive computer services from the liabilities that attach to other publishers or speakers, but this is a statutory decision that does not alter existing constitutional law.[85] Laws protecting the reputations of individuals existed at the time of the enactment of the First Amendment and have continued consistently to this day. The Constitution also authorized Congress to create legal protections for intellectual property, protections interpreted to persist alongside First Amendment rights. These laws defend interests that are in tension with absolute rights of speech, and the laws' longevity and vitality well demonstrate the contextual, balancing approach to judicial enforcement of the First Amendment, even with the more aggressive First Amendment judicial views that have developed recently.

False expression that harms a person's reputation can be regulated and punished by the government. Such expression can cause injuries to individuals who are the target and also hurt their relationships with others. This view, well understood by those who drafted the First Amendment, still holds even alongside expanding conceptions of First Amendment rights and their enforcement. Historically, false assertions and negative statements about an individual fell outside of First Amendment protection altogether. Among the "well-defined . . . classes of speech, the prevention and punishment of which have never been thought to raise any constitutional problem" are expressions that are "lewd or obscene, the profane, the libelous, and the insulting or 'fighting words.'"[86]

Even as the Supreme Court has limited lawful defamation actions in the context of public figures, it has treated written and spoken slanders against individuals as a basis for monetary damages and other remedies. So in 1964 the Court concluded that negligence should not give rise to a libel action against public officials; later this was extended to individuals involved in matters of justifiable and important public interest, but it did not foreclose actions for defamation in these or other instances.[87] By requiring a higher standard of proof (that is, more than mere mistake) for libel actions brought by public officials, the Court sought breathing space for criticism of those in government and public affairs, and worked to avoid deterring speech involving public officials and public figures.[88] Nonetheless, here and in subsequent matters, the Supreme Court has concluded that the traditional libel and defamation laws and the First Amendment could and should accommodate one another, even if false claims can sometimes secure constitutional protection.[89]

Over time, the courts altered long-standing common-law rules governing defamation through new interpretations of the First Amendment. The Supreme Court ruled more than fifty years ago that government regulations forbid speech expressing group-based biases or hate. And although this decision collides with subsequent judicial reasoning given the more libertarian turn in the courts' to speech, the Supreme Court has still not offered First Amendment protection for speech that is both false and harmful to a group's reputation.[90] Some other courts and most commentators suggest that group libel laws cannot survive other developments in First Amendment law, but others defend the Supreme Court's precedent for permitting such laws, and it has never been overruled.[91] Defamation may be actionable even without naming a specific person if negative comments about a small group are reasonably understood to refer to those individuals who object.[92] The rise of internet blogs and websites allowing posting by anyone—not just journalists overseen by to human editors—raises potential questions about the scope and application of such laws against group libel, but laws defining and providing sanctions against libel and defamation continue to hold force. Recognizing both the justified effort to protect individuals against private injuries to their reputation and the need for journalism to have room for expression, the Supreme Court ruled, "We hold that, so long as they do not impose liability without fault, the States may define for themselves the appropriate standard of liability for a publisher or broadcaster of defamatory falsehood injurious to a private individual."[93]

Similarly, governmental protections of intellectual property are permitted despite tensions with the First Amendment.[94] When the Supreme Court rejected First Amendment claims to fair use of a substantial portion of the then-unpublished memoirs of President Gerald Ford, the Court recognized and protected the right to first publication established by the federal copyright law.[95] Individual cases may articulate more or less room for intellectual property protections versus expression of the rights of those who did not originate the material; here as elsewhere, the First Amendment has not meant "no regulation" restricting speech.[96] Writers and publishers who republish the original material of others without permission or compensation violate the copyright laws and can be punished accordingly unless a specific exception applies.[97] The same rules apply to online expression, with certain statutory exceptions for hosting, storing, or transmitting third-party materials.[98] Because U.S. law implements international treaties, changing treaties to eliminate exceptions would be difficult. And despite tensions with what purely domestic U.S. law might call for, the law exposes internet platform companies and internet service providers to liability for their own defamatory conduct and direct or collusive infringement on intellectual property. Amid First Amendment rights, the persistence of libel and copyright laws attaches to internet speech and points to avenues for new laws.

Antitrust and Other Structural Regulations

The First Amendment's directive that Congress "shall make no law . . . abridging the freedom of speech, or of the press" does not prevent Congress or the executive branch from ensuring competition and guarding against negative effects of concentrated ownership and power in the media. Authorized by the Constitution to regulate interstate commerce, the federal government has devised laws and implemented rules restricting concentrated ownership and anticompetitive behavior by media companies. Doing so emphasizes that economic and technological factors affecting who has access to and control of communication media give rise to obligations on those with such power to protect the interests of the public. The economic and governance structures of media remain subject to general laws addressing market allocation, bid rigging, price fixing, monopoly and oligopoly market dominance, and other unfair conduct-restraining trade that interferes with fair competition or tends to produce unfair or deceptive practices where the effect on speech is incidental or indirect.[99] The application of these goals necessarily varies in different market contexts. Prohibiting ownership by the same entity of a broadcasting license and a newspaper in the same community made sense as a measure to advance competition and promote diverse views in an era before the alternatives enabled by cable, satellite, and the internet.[100]

In 1945, the Supreme Court approved antitrust enforcement despite the asserted defense that the First Amendment barred such legal action.[101] The defendant, the Associated Press, was (and still is) a cooperative service, which, as noted in Chapter 2, is permitted by antitrust law. At the time of the suit, the bylaws of the AP and its contracts with members prohibited service of AP news to nonmembers, barred members from supplying news to nonmembers, and authorized members to block competitors from becoming members.

These contract terms deployed concentrated power to operate as an illegal restraint on trade, reasoned the Court, and freedom of the press from governmental interference under the First Amendment does not allow private actors to repress the freedom of others.

The Supreme Court endorsed and reinforced this reasoning in *Miami Herald Publishing Co. v. Knight Newspapers*, even while finding defects in the particular Florida law at issue in that case, which required equal space for political candidates who did not receive a newspaper's endorsement.[102] Here the Court was especially vigilant, as the governmental focus on speech was direct. The Supreme Court pointed to its prior decision in the Associated Press case to distinguish acceptable policies for addressing the volume and quality of coverage from forbidden governmental acts suppressing speech.[103] Accordingly, the First Amendment is not a shield against otherwise lawful government regulation of private concentrations of power that restrict whose voices can be heard. The Constitution, according to the Supreme Court, is compatible with enforcement of antitrust rules in the context of media (small media entities can be exempted from labor laws) and with rules about ownership of different kinds of media in the same market.[104] Similarly, rules requiring print publishers to periodically report the names and addresses of their editors, publishers, and owners have been treated as compatible with the Constitution and the public purposes of transparency and accountability.[105] First Amendment theorist Alexander Meiklejohn put the point broadly back in 1948 by connecting freedom of expression to democratic self-governance: "What is essential is not that everyone shall speak, but that everything worth saying shall be said."[106]

The Supreme Court powerfully expressed concerns about concentrated economic control of media in the 1970s. The Court has observed how economic factors that shrink the number of metropolitan newspapers similarly make the entrance of new players difficult.

In this light, reasoned the Court, the First Amendment "acts as a sword as well as a shield" and "imposes obligations on the owners of the press in addition to protecting the press from government regulation."[107] In allowing antitrust regulation of media companies, the Court reasoned that

> the First Amendment, far from providing an argument against application of the Sherman [antitrust] Act, here provides powerful reasons to the contrary. That Amendment rests on the assumption that the widest possible dissemination of information from diverse and antagonistic sources is essential to the welfare of the public, that a free press is a condition of a free society. Surely a command that the government itself shall not impede the free flow of ideas does not afford nongovernmental combinations a refuge if they impose restraints upon that constitutionally guaranteed freedom.[108]

Abuse of market power warrants antitrust enforcement; this basic guiding principle of structural regulation continues to hold even if a defendant claims the First Amendment protects abuse of the power enabled by economic concentration.[109]

At the same time, not every regulation can overcome First Amendment concerns, as evidenced by the judgment that government cannot force a media company to afford a "right of reply" triggered by the company's own choice to include critical perspectives.[110] In the case of *Miami Herald v. Tornillo*, the law in question intruded too directly on the editorial voice of the publisher. Still, some regulation of editorial choices have withstood legal objections: the Court subsequently rejected a challenge to the requirement that cable companies carry local and public service content meant to advance the public interest.[111] In *Turner Broadcasting System, Inc., v. FCC*, the Court reasoned that this "must-carry" requirement was not a penalty imposed on the companies' choices about what else to cover and would not risk suppressing information or coverage of controversies.

The government's authority to challenge concentrated ownership and control of media and to advance the public's interest in receiving information could be exercised now and in the future to tackle concentration of control over the internet and its supporting infrastructure. The federal government's own involvement in developing and funding the internet could also supply grounds or even an obligation for government action to improve reliable access to material enabling competing views and authentication of messages and sources. Governing action aiming to alter the structure of internet entities falls within the ambit of permitted action, and may even be obligatory if the government itself is responsible for the designs that undermine the public's ability to obtain information needed for self-governance.[112]

Debate over whether the Constitution forbids, permits, or even requires structural regulation of internet-related activities exploded in the past decade around "net neutrality."[113] Net neutrality (the term was coined by law professor Tim Wu) is the notion that all internet traffic should be treated equally and without discrimination; therefore, internet service providers (such as Comcast and Verizon) should not be able to speed up service for some and slow down or even block it for others, whether based on willingness to pay, type of equipment, address of the source or destination, method of communication, user, content, or application.[114] Net neutrality drew the ire of many companies and the avid support of many nonprofit and free speech organizations. Net neutrality became law through an FCC regulation, which was upheld over legal challenge.[115] And then the Republican majority on the commission appointed by President Trump revoked it.[116] A further court challenge failed, except with regard to the argument that states should be allowed to adopt net neutrality requirements, and the debate continues.

The issue is likely to reemerge with a new administration and any actions by the government toward net neutrality requirements will undoubtedly face court challenges.[117] So it is worth noting the debate over the First Amendment in the context of the net neutrality rule.[118] In his dissent from the denial of rehearing in the challenge to the net neutrality rule, then–D.C. Circuit judge Brett Kavanaugh reasoned:

> Internet service providers and cable operators perform the same kinds of functions in their respective networks. Just like cable operators, Internet service providers deliver content to consumers. Internet service providers may not necessarily generate much content of their own, but they may decide what content they will transmit, just as cable operators decide what content they will transmit. Deciding whether and how to transmit ESPN and deciding whether and how to transmit ESPN.com are not meaningfully different for First Amendment purposes.[119]

Another judge on the panel, Sri Srinivasan, concurred in the decision not to rehear the case, and his opinion responded to the "misconceived" idea that the First Amendment entitles an internet service provider (ISP) to block its subscribers from accessing certain internet content based on the ISP's own preferences, "even if the ISP has held itself out as offering its customers an indiscriminate pathway to Internet content of their own—not the ISP's—choosing."[120] Judge Srinivasan (now Chief Judge) explained that an ISP has no First Amendment right to use such practices because

> the First Amendment does not give an ISP the right to present itself as affording a neutral, indiscriminate pathway but then conduct itself otherwise. The FCC's order requires ISPs to act in accordance with their customers' legitimate expectations. Nothing

in the First Amendment stands in the way of establishing such a requirement in the form of the net neutrality rule.[121]

Another structural regulatory approach would deem the internet a public utility, subject to regulation. Whether such an approach is compatible with the First Amendment, however, is a new problem. Public utilities in the past have not been entities that regularly control and select messages to distribute to others, although the short-lived net neutrality regulations went down that road before a change in political regimes intervened.[122] In addition, with AT&T's acquisition of Time Warner and CNN, the distinction between companies that carry messages through wires and those that create and share speech dissolves.[123] As state and federal officials explore the possibilities of public utility treatment for internet service providers and other tech companies, the issue will come up for debate.[124] What parts of telecommunications have been treated as a public utility, subject to regulation, could very well change to include internet access.[125] The Supreme Court has shown its ability to protect the First Amendment rights of a public utility: the Court accorded constitutional protection to the Pacific Gas and Electric Company's own choices about what and where to speak, and allowed it to refuse to distribute to users a rebuttal to its own message.[126]

At issue—and unresolved—is whether the First Amendment can be invoked to shield the companies providing cable and telephone infrastructure from a wide variety of regulations, an issue that arises because of the libertarian view that turns the First Amendment into an all-purpose tool against government regulation. As technological and business practices could entrench further a situation in which a few major players govern access to and structure of internet-based communications, the public needs protection, and hence the government needs some latitude for action. Recognizing the significance of internet settings for the expression and reception of speech by providers and by the public should not foreclose regulation of economic concentration. Nor should the First Amendment preclude legal oversight of a resource that has become a necessity of life. Government tools of taxation and subsidy—authorized by the Constitution's tax and spending clause—remain available unless they are deployed to single out some speech or other content.[127] As is true with other media, contextual tailoring to the nature of government interests and the scope of government measures permits government action to remain consistent with the First Amendment.

Taxes and Subsidies

The government may impose taxes on the press or media but may not do so in ways that regulate content or that single out the press or a subgroup of the press for differential treatment. Even content-neutral laws require heightened scrutiny—placing the burden on the government to demonstrate how the tax is narrowly tailored to serve a compelling interest—if the tax targets or threatens to suppress particular speakers or ideas, or if it singles out the press or another part of the media.[128] Yet a tax that falls more heavily on some forms of media rather than others can be upheld if it does not discriminate on the basis of ideas and instead is a general tax, not one selecting a narrow group to bear the burden.[129]

The foundational decision *Grosjean v. American Press Co.* rejected a license tax based on the size of the newspaper circulation with the recognition that the larger publications bearing the tax burden were likely to be critical of the state's senator (and former governor), Huey Long.[130] The apparent intention to impose the burden on a small group of newspapers and to limit the circulation of those papers' ideas proved critical. Similarly, in a later case, the Court rejected a special use tax on ink and paper as a violation of the First Amendment because it targeted both a small group of newspapers and the press itself. Extending the same approach, a tax exempting "religious, professional, trade and sports journals" but no other types of publications was found to be an impermissible regulation of content.[131] The press and media are not constitutionally immune from general taxation; the heightened judicial scrutiny arises only where the scheme treats them differently than others or selects among types of media outlets.[132]

Yet the Court found no defect in a scheme providing an exemption from sales taxes for newspapers, magazines, and direct satellite broadcasting but not for cable television services. With cable services being numerous and offering a variety of programs, this differentiation did not even give rise to heightened judicial scrutiny in the 1991 case *Leathers v. Medlock*.[133] The Supreme Court distinguished defective tax schemes that had targeted a limited group of speakers and thus posed a risk of censorship or constricting views or ideas. Treating different media differently is not a constitutional problem unless it implies discrimination on the basis of ideas and views.

On this basis, even if internet platform companies are viewed for some purposes as publishers, they could be taxed differently than other media or news organizations. Similarly, a tax scheme that treats not-for-profit organizations differently than ones established as for-profits would not be a problem unless it could be shown that such a distinction had an intent or effect on content or viewpoints available to the public. For-profit organizations may well serve public purposes, but they are established to make a profit to be distributed to owners. Nonprofit organizations are established to advance a public purpose, and any revenues that are generated stay within the organization. Economists would say that the exemption from taxation (and deductibility of donations) amounts to a public subsidy assisting the enterprise. When a nonprofit organization avoids a tax, it is saving money that a competing for-profit organization has to pay. But can the insulation from a tax include a condition that the nonprofit organization has to give up some of its First Amendment freedoms, such as freedom to take political positions? No, said the Supreme Court.[134] Unclear, however, is the line between impermissible conditions on a public subsidy and permissible government payment for speech carried by a private entity.

The Supreme Court has concluded that a public subsidy is compatible with the First Amendment even if its scope is vague because when the government acts as a patron rather than as a sovereign, the consequences of its decisions are less significant constitutionally.[135] The award of public dollars may include conditions limiting access to certain knowledge and materials: in *United States v. American Library Association* (2003), the Court upheld a statutory provision requiring schools and libraries receiving public funds for computers and internet access to filter out internet material considered objectionable for children.[136] The Court reasoned that libraries are not public forums and that the required filters would be analogous to selection of books by librarians, although filters both under-block and over-block.[137]

The Court's treatment of public subsidies and speech may produce odd results. The Court allowed the government to direct that health care providers taking government funds may not inform their patients about the availability of abortion, but it rejected as unconstitutional a directive forbidding publicly funded legal services attorneys from representing clients challenging government welfare laws because pursuing such litigation is itself a form of protected speech.[138] Scholars question judicial precedents that allow the government to pick and choose among the messages it subsidizes.[139] Yet neither scholarly treatments nor judicial doctrinal analyses have prevented the government from subsidizing media and leaving editorial and content decisions to those who manage the subsidized entity. Indeed, the First Amendment forbids viewpoint discrimination in the use of public funds for student journalism. Nor can the government ban editorials by publicly funded broadcasters.[140] First Amendment analyses depend on context, including the nature of each medium and the public policies at stake.

Voluntary Private Action Versus State Action

A basic building block of First Amendment analysis preserves private companies' expansive power to regulate speech. This building block for analyzing a constitutional problem is the "state action" doctrine. The Supreme Court views the essential trip wire for First Amendment review to be actual governmental action affecting speech. To satisfy this requirement, the action in question usually must come from a government official, a legislature, a city council, or a public university.[141] The First Amendment itself indicates that it is Congress that must not make a law abridging the freedom of speech. In addition, the Supreme Court has interpreted the First Amendment to apply not only to the federal government but also to the states through the Fourteenth Amendment's prohibition of state action depriving a person of life, liberty, or property, satisfied that freedom of speech is such a fundamental liberty. These constitutional requirements generally do not attach when the action in question comes from a private individual or company.[142]

Because it is the gateway to constitutional protection, the "state action" doctrine has become an undeniable mess, with inconsistent and unpredictable results.[143] Even a private actor can be found to satisfy the state action requirement under some circumstances, such as deep entanglement with government through contractual or property relationships, through taking on essentially public functions (the chief example involved a company town), or through reliance on the government to enforce a private arrangement.[144] Unlike other nations that extend constitutional requirements to the conduct of private actors, the state action requirement in the United States embodies judicial concerns about private actors' freedom and reflects particular efforts to define the roles of state and federal governments, courts and legislatures, and economic markets.[145]

The internet and other digital communications—and the companies they enable—have such significance in people's lives that they begin to feel like public streets, broadcasting, or government. Individuals, families, schools, businesses, governments, and other sectors are increasingly dependent on the internet and other digital platforms for communicating, storing, and retrieving information; assessing conduct and performance; conducting work and schooling; and even resolving disputes. The state action issue determines whether internet platform companies are to be bound by the First Amendment (or, for that matter, by constitutional norms governing individual privacy and equal protection). For constitutional purposes, might an internet platform company function like a company town, and require the platform company's "streets" to be open to free speech just as the company town's physical streets had to be open to free travel? Federal district court judge Lucy Koh dismissed a case alleging unconstitutional censorship by YouTube when it removed video uploaded by a conservative private college.[146] The reviewing court agreed, observing, "Despite YouTube's ubiquity and its role as a public facing platform, it remains a private forum, not a public forum subject to judicial scrutiny under the First Amendment."[147] Even though internet platforms may be taking the place of public streets and parks when it comes to debate and discussion, the constitutional analysis identifying state action does not apply; hosting conversations is not a function reserved to government. YouTube is a private entity operating privately.

As this case indicates, what is "public" and even what reflects the involvement of government do not determine what counts as state action for purposes of constitutional guarantees. The open accessibility of internet platforms makes them seem public in the sense of available to all and communal. They are not public in the sense of paid for by government funds or ruled by government officials or employees. Yet conceiving of internet platform companies as entirely private obscures the degree of government involvement in their existence and operations, which goes well beyond mere licensing and regulation. The sheer fact of government involvement in the creation of the internet and the algorithms used by companies such as Google does not by itself amount to "state action."

Hence, the actions of internet companies do not trigger constitutional requirements of freedom of speech, due process, and equal protection. But both historical governmental involvement and the dependence of the entire political system of representative democracy on a viable news industry weigh in favor of reforms. Reforms both in the realm of law and reforms that are initiated within the private domains of companies, advocacy organizations, and nonprofit alternatives are needed to address public interest concerns about the quality of available information, fairness in what messages are promoted or demoted, and treatment of hateful or conspiracy expression. Thus, even if an action by a given internet or tech company is not "state action," the company can choose to embrace constitutional values as part of its own identity and commitments, or it can pursue other values, such as creating a hate-free environment. Private companies, including private publishers, private schools, and religious groups, can assert their freedom of expression without the restrictions that attach to federal, state, or local government. For example, Twitter can ban hate speech and remove it when posted and also can decide to permanently ban a public official for violating the company's policies against glorifying violence.[148] Facebook can remove ads purchased by a presidential candidate for violating its own policies against hateful messages but also shut down efforts to make the site less divisive and later wrestled with allowing President Trump's posts until their association with the violent insurrection at the United States Capitol on January 6, 2021.[149]

Internet platforms are allowed to make choices about what priorities guide results from internet search algorithms; such choices have been given constitutional protection.[150] As the platform companies learned during the COVID-19 crisis, aggressive and publicly disclosed efforts to edit, remove, and prevent misinformation are in their interest. Even before the pandemic, Twitter reported that its "long-term success depends on [its] ability to improve the health of the public conversation on Twitter" by taking down spam, fake accounts, hate speech, and other faulty content.[151]

But as companies in ongoing relationships with consumers and users, they also have obligations. Internet platforms can adopt mission statements and make enforceable contractual promises to their customers and users; they can also govern themselves by adhering voluntarily to constitutional norms. Consumer protection enforcement at both state and federal levels can bolster commitments by the platforms, and state law can protect against breach of contract.[152] The legality of contracts and fair-dealing requirements is straightforward. Contracts with customers to provide services are enforceable against internet platforms. A federal court rejected Google's effort to dismiss a complaint alleging unlawful, unfair, and fraudulent conduct under California's Business and Professions Code after plaintiff Dreamstime—a seller of stock photos—found itself down-ranked in Google search results, search advertisements, and mobile applications.[29] Internet platforms do not evade duties of care even if a user has no contract and receives services for free (actually, in exchange for their personal data). Class action plaintiffs alleged that Yahoo failed to protect their sensitive data, and a federal court found plausible a claim that Yahoo behaved negligently, failing to make timely disclosures; furthermore, the company could not enforce the liability limitations specified in its terms-of-service agreement, as they were "substantively unconscionable" elements.[153]

Platform companies produce rules for their own conduct that could be enforced, and governments may make and enforce general consumer protection requirements—subject to constitutional restraints.[154] A terms-of-service agreement may include an enforceable duty to disclose breaches of security in handling user data.[155] An internet platform may rely on courts to uphold its terms-of-service agreement. AOL successfully turned to both state and federal courts to defend against a First Amendment challenge to its use of spam filters for mass mailings in *Cyber Promotions v. America Online*.[156]

CONSTITUTIONAL ARGUMENTS FOR THE AFFIRMATIVE DEMANDS OF THE FIRST AMENDMENT

Because the Constitution depends on informed and active members to make the democracy it establishes work, the Constitution should compel development of the institutional context for democratic self-governance. Larry Kramer, former dean of Stanford Law School, notes, "You cannot run a democratic system unless you have a well-informed public, or a public prepared to defer to well-informed elites."[157] He warns of the dangers from failures by Google and Facebook to engage in fact-checking, and notes they are inevitably selecting material; he argues that they have obligations to do so in a way that supports democracy's prerequisites.[158] Affirmative dimensions of the First Amendment include the rights of listeners and attention to diversity in participants and in ideas needed by a democratic system.[159] As the Supreme Court has observed, "It would be strange indeed . . . if the grave concern for freedom of the press which prompted adoption of the First Amendment should be read as a command that the government was without power to protect that freedom."[160]

Government action always carries risks and needs to comport with constitutional guarantees, but government inaction can also jeopardize constitutional guarantees. If the economy collapses, the government takes action.[161] If the basic mechanisms for collecting votes become vulnerable to hacking, the government should act.[162] And if the infrastructure for gathering, reporting, and distributing news is absent in many communities, if readers and viewers are overwhelmed by distractions designed to take their attention, and if no recourse is available through the accountability mechanisms designed for either government or private enterprises, it is time to return to the Constitution's text and basic principles.

Take, for example, the confusion arising when government officials rely on social media platform companies for communicating with the public. Once again, the state action doctrine matters, but current practices fall outside anything imagined by those who initially articulated it. Site architecture allows government officials using private tools to constrain speech and the press; they are officials controlling who can speak with them, but they are doing so through the mechanisms provided by private companies. According to the platform policies, the president of the United States can block particular individuals from receiving his messages and from communicating back by using the functionalities afforded by Twitter or Facebook. Conduct by private actors eludes First Amendment consideration, as only when local, state, or federal government abridges speech or press have the courts found constitutional violations. But when a public official announces policies and views through a private platform—sometimes exclusively that way—then the traditional lines between public and private allow abuses that undermine democratic governance. The public official's communications should be viewed as a kind of public communication, so they should not be able to block or "unfriend" individuals seeking to connect with them.[163] It is not that Twitter is acting as a government, but it is providing a platform that a government official has converted into a limited-purpose forum by using it for governmental communications. Government officials who convert private vehicles into their official channels of communication should be subject to the Constitution. Digital companies in these and other contexts could be seen as functioning like governments, controlling the public squares where people communicate. Under these circumstances, how should the First Amendment apply to internet platform companies?

It may seem as though the next right legal step is to define all of social media or all internet platform companies as the equivalent of public streets and parks, held in the public and treated as governmental for purposes of the First Amendment. Indeed, Justice Anthony Kennedy raised this possibility in observing for the Supreme Court that people may be communicating more online than in public parks and streets: "While in the past there may have been difficulty in identifying the most important places (in a spatial sense) for the exchange of views, today the answer is clear. It is cyberspace—the 'vast democratic forums of the Internet' in general, and social media in particular."[164]

Deeming all of cyberspace as "public" for purposes of the First Amendment could seem to follow the precedent of *Marsh v. Alabama.* There the United States Supreme Court applied the First Amendment to a town owned by a private company and ruled against using a law banning trespassing on private property against a person distributing religious leaflets.[165] The Court reached this unusual result because the streets looked like streets in an ordinary town, because the town was relatively open to the public, and because the rights of individuals to enjoy freedom of speech and of religion hold a privileged place even against the rights of property owners. Internet platform companies and social media networks have some of these qualities, including being open to the general public.

Nonetheless, extending the reasoning of *Marsh* to internet platform companies is both unlikely and unwise. It is unlikely because expansion to any private company would eliminate the state action requirement, and the Supreme Court shows no appetite for so doing.[166] Yes, just as in a company town, people may confuse internet channels of communication with public ones and object to censorship of their speech there, but people really do have avenues for expression outside a particular internet platform.

And it is unwise because if we view internet platforms as functioning like the government, their decisions to remove content or alter access by others would be subject to First Amendment challenges, meaning that private platforms would be barred from guarding against harassment, bullying, and deceit—including misinformation about COVID-19 and hateful or violent materials. Twitter, for example, could not ban high-profile white supremacists, and Facebook could not remove information that violates its community guidelines. Applying the First Amendment to a platform company blocks it from trying out methods of moderation such as algorithmic machine learning, because such systems make errors while they are being trained.[167] Governance is hard, as scrutiny of self-regulation by the digital platforms shows.[168]

Similarly, it would be problematic to treat each action by an internet platform company as speech if that would erect a First Amendment shield against regulation of fraud, deceit, or manipulation—as when a platform tries to alter the mood of its users without notice or consent.[169] Using the First Amendment to strike down regulation of internet platform companies could prevent contract enforcement, obstruct antitrust regulation designed to counter economic concentration, or ban required disclosure of the source of political advertising even if the source is a terrorist organization or a foreign government.

Certain institutions and practices are necessary for democracy to proceed.[170] Recent experiments with new democracies that have fallen into tyranny underscore this lesson.[171] Because reliable circulation of actual news is one of those crucial institutional practices, the community, and indeed the government, should not sit idly by and watch the news industry collapse.[172] If the Constitution is "not a suicide pact," then the Constitution should not stand in the way of measures to revitalize the news media. The First Amendment's presumption of an existing press may even support an affirmative obligation on the government to undertake reforms and regulations to ensure the viability of a news ecosystem.[173] This notion of a positive First Amendment, developed repeatedly by scholars and commissions, appears in the reasoning and results of some judicial decisions and deserves recognition and action in light of the demands of democracy under serious stress.[174]

Different constellations of economic, social, and institutional relationships make democracy more or less possible. To work, democracy needs (1) an arena where participants can engage in self-governance; (2) institutions enabling individuals to learn about social needs and personal desires, to deliberate, to express their views, and to select representatives to do the work of governing; and (3) the circulation of information that enables people to act to advance their own and society's interests.[175] These predicates are in jeopardy. Federal action is needed to guard against overconcentration of economic power, which shields digital companies from competition, accountability, and fair contributions to news media. Concentrated power allows internet service providers to skip rural and small communities in their infrastructure projects or forestall public obligations to meet those needs.[176] Federal action is necessary to overcome local news deserts, which jeopardize the health of residents, the effectiveness and efficiency of local governments, and the ability of residents to engage meaningfully in self-governance. News deserts leave many local communities without reporting on local governmental and community developments, denying many people what they need to govern themselves and hold others accountable.

Intellectual property protection and enforcement, expressly authorized by the Constitution, can ensure compensation for the work of journalists that is at risk of appropriation by third parties posting on an internet site. It requires federal action, as this is a body of federal law. Digital companies free ride on the news links shared by users without reinvesting in the apparatus necessary for investigating, testing, and reporting news, which undermines people's ability to get and trust news. Federal action would be important to combat security defects in digital communications that seem to permit foreign manipulation of campaign speech. An architecture of online communications reliant on algorithmic moderators enables anonymous and bot-initiated messages to populate the news that individuals receive. Anonymity for speakers can protect against intimidation and harassment, but those values need to be weighed against interests in integrity in political campaigns, liability for legal violations, and national security.

First Amendment freedoms and the crucial watchdog function of news do not hinge solely on the viability of news-producing enterprises. They also depend on redressing damaging decisions (made by dispersed, powerful, private, competing companies that are not in the news business) that are invisible to those affected and that may be demonstrably manipulated by enemies of the nation and treacherous schemers. To sustain freedom of the press and enable democratic self-governance, courts should pursue new approaches, and so should Congress. The Constitution can and should play a vital role in exposing how the government contributed to the dangers we currently face and should therefore prompt new policies, regulations, and practices.

The federal government could be understood as having a moral and constitutional duty to take action—legislatively and through the tools of law enforcement—to protect the generation, production, and distribution of news. These are ideas of an affirmative First Amendment stand beyond current law, but today's challenges to the very premises of constitutional democracy warrant big and serious ideas. These include a hard look at the development of First Amendment doctrine and an interpretative guide to make sure it does not produce unnecessary barriers to the government actions needed to revitalize the news business.

The Constitution is implicated because the democracy it establishes depends on a citizenry informed by news in ways that are currently in severe jeopardy. The First Amendment in particular is implicated, because it was meant to protect the right of readers and listeners to gain access to information as well as the right of speakers and the press to express themselves. The First Amendment has never barred all government action that affects speech; government action shapes the internet and media in specific ways that contribute to the very jeopardy of the current times. The complex ecosystem of contemporary media did not spring solely from private decisions, investments, or market strategies; it has been shaped, supported, and promoted by specific government policies and actions. At a minimum, First Amendment values strongly support the policy of government action here; inaction by the government amounts to failures to fulfill constitutional obligations.

CHAPTER FOUR

. . .

Constitutionally Inflected Reforms

WHEN IT COMES to saving the news in today's media ecosystem, what we have is a "wicked problem," which means a problem with multiple interacting causes and no single solution.[1] No one initiative would be sufficient to address the forces that have changed the production, distribution, and reception of news over the past several decades. The introduction of digital communications led to the migration of attention, advertising, and financial investment to a few global tech companies that do not consider themselves to be in the publishing industry, much less the news businesses. Legal rules and practices insulate those companies from the liability publishers face and from government regulation to advance consumer protection, restrict monopolies, and oversee public utilities. And although manipulation of human attention and beliefs has always been possible, digital tools enable both sophisticated and mercenary actors. Digital tools amplify such efforts at manipulation with little cost and produce enormous effects on what people think they know and trust. The problems are compounded by government failures to enforce existing laws and by the libertarian weaponizing of the First

Amendment. Protecting and advancing the constitutional promises of freedom of speech, freedom of the press, and a republican form of government require initiatives addressing the financing, infrastructure, variety, and accountability of media.

One cluster of issues has to do with the destructive effects of major internet companies. Their business models, operations, and insulation from liability and law enforcement disrupt older media and news industries without contributing resources to news producers upon whose content they free ride. Too many public and private actors fail to respect and enforce existing legal rights and protections concerning deception and exploitation. Advertising dollars migrate to digital resources—including the shift from classified ads to online sites—and decimate the revenue base for local journalism. Digital platform companies in turn gather data on users, target ads, and seek to maximize traffic on internet sites—amplifying false and hateful content. Legacy media companies made mistakes such as offering content for free and initially deciding not to enforce their copyright claims. These trends taken together give rise to growing distrust of information and of those who purvey it.

A second set of issues reflects the political and legal shifts in the United States away from public interest regulation and toward libertarian, free market, and pro-business policies—with resulting explosion of economic inequality and its companion, political populism of both left- and right-wing varieties. Hollowed out are public enforcement against concentrated economic power and consumer exploitation. Missing are commitments to First Amendment law that advances the rights and needs of listeners, readers, and citizens and finds affirmative government duties to ensure an informed populace.[2] An older vision—manifested in the postal system and public media subsidies, the government regulation of telecommunications and economic concentration, and the aspirations of public service—has deep historical roots and rich philosophic defenses that could be wellsprings for needed reforms.

The Constitution's framers did not anticipate this combination of technological and political changes that expose the limitations of relying on for-profit news businesses to ensure the freedom of speech and freedom of the press vital to a functioning democracy. Compounding the difficulties are a ballooning distrust of media, government, facts, and law as well as the growing complexity of technology and government. Such complexities make it difficult for even the most highly motivated individuals and civic organizations to track and push back against troubling practices of internet platform companies or to guarantee sustainable sources of local news. In the face of news deserts, growing concerns about hoaxes and misinformation online, and increasing distrust across society, major digital platform companies know they have a legitimacy problem. Facebook's creation of an oversight board and "verification badges" for public figures on its subsidiaries Instagram and WhatsApp are examples of private sector responses; so are periodic announcements of new rules banning election disinformation, Holocaust denial, and even a sitting U.S. president.[3] Yet even these efforts generate new debate and, potentially, further distrust. Although no single reform will fix the decline of news industries, the global rise in distrust of information sources, and the wasting away of America's public interest commitments, three types of responses could help. The first type would treat digital platform companies as responsible players, subject to duties and expectations commensurate with their functions and their powers. A second kind would vitalize public and private protections against deception, fraud, and manipulation and bolstering the capacities of individuals and communities to monitor and correct abuses and demand better media and internet practices. A third category of changes would support, amplify, and sustain a variety of public interest news sources and resources at the local, regional, and national levels. Taken together, these are steps that would halt and might reverse the destruction of journalistic news—involving the platform companies in sharing their gains to reinvest in news development and paying for the content they take—and also renewing and strengthening the public service and public interest dimensions of law enforcement and in media investments. Here is a sketch of specific examples—twelve in all—illustrating each category of initiative. The hope is that these ideas will spur more debate and generate improved options to tackle this wicked problem.

TREAT INTERNET PLATFORM COMPANIES AS RESPONSIBLE ACTORS

Internet platform companies may once have been fledgling innovations, but now they are among the largest and most financially successful companies in the world. For this reason, and in light of their effect on the news industry and democracy, it is time to treat them as responsible for what they do, commensurately with their functions and powers, and insulated no longer from the duties and expectations that apply to others.

Require Payment for News Circulated on Social Media

Currently, much of the news people read is shared for free on social media, even when it is from sources protected by copyright. Recent studies show that people trust news coming to them from friends even more than they trust traditional journalism sources.[4] Digital platforms report to investors and advertisers the size and demographics of their audiences and take in ad money accordingly.

The journalists writing the content and their newspapers and broadcasters, however, do not share, other than in limited ways, in the platforms companies make deploying this content. Benefiting from the "safe harbor" clause in the 1998 Digital Millennium Copyright Act, Facebook and Google free ride on the content others create.[5] This clause could be changed to ensure that journalists and legacy news organizations are compensated for their work.[6] Platforms would need to pay; they could pass the costs on in the prices they charge for ads and other services, but they should bear more of the actual costs, especially as they deploy traditional news stories to bolster their users' trust.[7]

Even a modest requirement that digital platforms nudge their users to contribute voluntarily to news sources would generate resources to support investigations, editing, checking, and reporting. Framing matters. As a marketing expert put it: "Why would people who think nothing of paying $5 for a Starbucks latte believe that a $10-a-month music-streaming service is overpriced?"[8] It took time for iTunes and Spotify pricing, streaming, and copyright enforcement to break music piracy, but it largely worked.[9] A similar combination of smart pricing, technological innovation, and regulation could restore income streams for journalism.

Payments for content may emerge as a competitive strategy. Facebook and Google have started to offer compensation for some stories from some outlets.[10] The actual payments thus far are not significant and do not reach local publishers. The efforts by Facebook and Twitter to fill the deserts of local news with updates about events and emergencies offer some needed information, but they focus on breaking news, like road closures or active shooters, and not on comprehensive or investigative journalism.[11] Pursuing more high-quality content, Apple pays to offer its users access to the *Wall Street Journal* and some magazines—and charges its users for these services.

Why leave that to the whims of the platform companies when the intellectual property and labor of those who produce the content and the actual sustainability of news gathering and reporting are at issue? Jurisdictions outside the United States are beginning to require payment to content producers. The Australian government instructed the chair of its Competition and Consumer Commission to require internet platforms to negotiate payments with newspaper publishers.[12] The European Union has adopted rules requiring search engines to pay for journalistic matter (anything more than very short extracts) used for a commercial purpose by an "information society service provider."[13] This directive, to be implemented by member nations, enforces copyright interests and should provide at least some resources to the owners of the copied materials.[14] France plans to enforce the directive. Facebook has already begun to give payments to publishers for licensing their content.

Devising compensation requirements gave viability to the music industry and its artists and contributors in the United States. The American Society of Composers, Authors and Publishers developed the concept of a "blanket license," under which businesses such as restaurants, radio stations, and retail stores gain the right to play a composition in exchange for a fixed annual fee, ensuring access revenue streams to which music contributed.[15] A similar analysis could applies to news reported by journalists and their employers that is posted onto digital platforms and draws users' and advertisers' attention.

Paying for content would provide revenues to journalists and their media organizations. There would be questions to be answered, of course: Who counts as a publisher entitled to compensation? How can some of the money reach actual producers of the content who may not be the rights holders? And what exemptions should exist for open-source materials, repositories, and data mining done by research and cultural institutions?[16] These are familiar issues for any situation involving intellectual property, and they are solvable.

Enforcing intellectual property rights for news means providing compensation to producers that would help sustain the reporting and writing of material that otherwise is at risk as conventional journalism organizations falter.

Some may worry that a compensation requirement would produce more paywalls or otherwise render inaccessible materials needed for the historical record and for research about past events. To mitigate such worries, the right to compensation could expire two years from the date of first publication. Will some internet providers prefer to withdraw rather than to pay for content and further reduce news available to the public? As an interesting initial clue to impact, an earlier Spanish law similar to the EU directive showed that although Google withdrew from Spain, internet traffic on Spanish news sites did not decrease and in some instances increased.[17]

Curtail Immunity of Internet Platforms and Subject Them to Liabilities That Attach to Traditional Publishers

Publishers in the United States are liable for publishing illegal or other legally actionable content; distributors such as booksellers, however, are usually not liable on the assumption that they are less likely to be aware of the content. As digital communications collapse the distinction between publishing and distributing, internet platforms could be treated as publishers. The 1996 Communications Decency Act's Section 230 exempts internet computer service providers from rules that apply to publishers, authors, or speakers of information, in an effort to promote innovation in what was once the new frontier—the internet.[18] Advocated by the tech industry and by many excited about the frontiers of freedom opened by the internet, Section 230 creates a hole in liability and removes incentives for responsible behavior by the platforms. With a few exceptions, this provision specifies that "no provider of an interactive computer service shall be treated as the publisher or speaker of any information provided by another content provider."[19]

The immunity offered by Section 230 allows private actors—such as interactive computer service providers—to filter, or not filter, objectionable content without concerns about legal challenges.[20] Meant to insulate the internet, which was still relatively new at the time, and the fledgling companies using it, Section 230 also represented a dream of a free space encouraging creativity. And it avoided debates over whether internet companies should be analogized to publishers or to distributors. Initially courts interpreted this provision to protect internet companies from liability for civil and criminal actions for facilitating sex trafficking until the law changed due to later court challenges and congressional modifications.[21] The section still protects internet platforms from liability for defamatory messages posted by a third party,[22] terrorist content developed by a third party,[23] false information,[24] and housing advertisements by third parties discriminating in violation of the Fair Housing Act.[25] Courts have also construed Section 230 as a shield against challenges to the sale of guns sold to people who cannot pass background checks.[26]

Although some still defend Section 230 as a spur to free speech, job creation, small business development, and consumer reviews, scholars and lawmakers from a variety of viewpoints and political positions in the United States have considered altering or eliminating Section 230 immunity.[27] The justifications offered for the immunity in the early days of the internet are less powerful now than they were in earlier decades. Internet-based businesses have grown to be among the largest and most successful businesses in the world. Step-by-step narrowing of Section 230 immunity has not produced serious problems. Conventional publishers in the United States face a disadvantage as long as their functional competitors online are immune. Section 230 immunity has allowed internet platforms to avoid responsibility for the disinformation, violence, cyber sexual harassment, and cyberbullying they propel.

Conservative critics have a different objection; they charge Section 230 allows the platform companies to muzzle conservative voices. When Twitter, in May of 2020, attached a fact-check response to a tweet by President Trump, President Trump retaliated with an executive order seeking to remove Section 230 immunities from tech companies that take action such as removing tweets or users without providing "a fair hearing."[28] The president actually lacks the power to do what he announced, and the status of Section 230 lies with Congress and with rulemaking efforts that are constrained by the First Amendment, statutory limits, and the participatory response required by administrative procedure. The right question for future reforms is whether tech companies now should simply face the same liabilities as publishers or distributors. Proposals working their way through the Department of Justice and Congress can more carefully address this complex question and the platform companies, perhaps reading writing on the wall, indicate openness and even eagerness for changes in the law.[29]

The internet platforms—offering access to millions of websites and many other digital services—do differ from publishers and producers that have to deal with scarcity of space on physical pages or airwaves, but even in its original version, Section 230 contemplated that the tech platforms would play an editorial role. Section 230(c)(2) grants immunity from civil liabilities when information service providers in good faith remove or submerge content that they deem "obscene, lewd, lascivious, filthy, excessively violent, harassing, or otherwise objectionable, whether or not such material is constitutionally protected."

Regulations from outside the United States govern global tech companies and prompt the companies to remove hateful and inciting materials. User groups have also pressured internet platforms to filter or block certain offensive content. Because internet platform companies from the start have used the room accorded by Section 230 to remove "objectionable" content, the question about their power to moderate and edit is not whether to use it but how. In fact, moderation is always happening, even when it is less than visible to users.[30] The companies use both people and algorithms to highlight, demote, or remove content. The major platforms acted with alacrity to remove conspiratorial or false information about COVID-19, demonstrating their moderation capacity. Subjects involving opinions call for more caution and care; historically, restrictions on speech have fallen more heavily on members of minority groups and also on those raising criticisms of existing power relationships. There are risks of error from digital moderation techniques relying solely on algorithms that neglect local and changing contexts, minority languages, or the disproportionate impact of criteria on minority groups.[31] But responsible publishers and editors know and attend to these kinds of risks and nonetheless find ways to advance accuracy and fairness.

The federal government can and should alter Section 230 because the rationale of shielding fledging new internet initiatives from any liabilities in order to promote innovation is no longer as necessary and contributes to unfair competition with traditional media. As a modest change, legal scholars Benjamin Wittes and Danielle Keats Citron propose allowing immunity from liabilities only to those internet platforms able to show reasonable content moderation practices addressing unlawful uses of its services that clearly create serious harms to others.[32] This proposal creates a content-neutral process requirement that begins to approximate the treatment of publishers elsewhere and would not violate the First Amendment. Immunity could be made conditioned on development by each company of internal plans to combat superspreading of disinformation. Nether such conditional immunities change nor the elimination of Section 230 altogether would jeopardize commitments to freedom of speech any more than liabilities imposed on traditional publishers do. Acting as private (not governmental) entities, the internet platform companies do not violate the First Amendment when they remove or down-list offensive or troubling content, and ending Section 230 does not alter this fact. If subject to legal liability—for libel, and even for inciting violence—the platform companies will have to step up.

Platform companies and search engines inevitably make choices determining what people see, and indeed they have First Amendment protections in doing so. Private internet companies could more transparently compete for users based on the qualities and elements of their moderation practices. Despite once denying they were in the journalism business, tech companies assert devotion to freedom of speech and raise it as a defense to a variety of challenges.[33] Congress announced in Section 230 that providers or interactive computer services would not "be treated as the publisher or speaker of any information provided by another content provider"; in this very language, Congress was acknowledging that otherwise the interactive computer services providers could be seen as publishers, speakers, or content providers. And indeed, this is a fair interpretation for companies that have and often use the power to exercise editorial functions—to choose what messages to present, to elevate, to make less visible, and to remove.[34] Some may have hoped that the platform companies would allow anyone to express anything, but that was not the original conception of Section 230 nor how the companies have proceeded (though the "dark web" hosts sites that avoid the review of platform companies). Companies that engage in editing, moderating, selecting, and deselecting—even by using algorithms—are not mere distributors, like a bookstore or newsstand. Those that prescreen or remove content have stepped into the moderating role. Moderating is a hard job and poses serious risks of mistakes and misjudgments.[35] That is why editorial judgments are accorded First Amendment protections. And that is why transparency about the methods used and results—with pressure to improve—matter. When Facebook mistakenly removed from its platform hundreds of anti-racist skinheads, including artists of color, critics exposed the error and the accounts were reinstated; the episode points to the need for better tools and processes.[36]

Internet platform companies should at a minimum face responsibilities and liabilities similar to those faced by distributors, which means liability for content they knew or should have known crosses the line into illegality. That test is met when users point out content that crosses the line. Where the platform companies more actively curate, select, and remove content, ending Section 230 immunity altogether is worth serious consideration, even though many claim it remains essential for internet communications. Debate over Section 230 is especially healthy if it prompts the platform companies, whether large or small, to be much more candid about what they do and how they do it—and face consequences in the marketplace, from critics, and perhaps from law reform as a result. Promising reforms would condition any immunity on federal agency oversight, and the platform companies themselves may prefer more clarity and shared responsibility for defining the right lines for content moderation.

As it stands, Section 230 amounts to a subsidy for a major competitor to legacy media organizations, and its practical and differential effects should be evaluated with no less scrutiny than the practical effects of other laws found to be impermissible abridgments of speech.[37] In the United States, courts grant tech platforms First Amendment protection for their editorial judgments, and the platforms could always defend decisions to include offensive or objectionable material on that ground.[38] Concerns that companies would adopt willful blindness and refuse all moderating functions seem unlikely given the reputational risks, consumer pressures, and regulations from outside the United States. Eliminating or substantially curtailing immunity of the internet companies would bring up to date the legal treatment commensurate with their power and responsibilities.

Antitrust Investigation and Possible Enforcement

Antitrust enforcement could address dangers and abuses from concentrated ownership and control that have insulated power-house companies such as Facebook and Google from competitive pressures, with resulting harms to consumer interests and barriers to alternative efforts offering differing content and terms. Recent developments in antitrust policies and doctrines have diminished concerns with market domination and consumer interests other than prices. The obvious remedy for antitrust violations, which is to break up Facebook, Amazon, Google, and other large tech companies into smaller pieces, may not help much. It is far from clear that breaking them up would do much to change the problems in the news ecosystem. Nonetheless, the threat of antitrust enforcement itself could lead to more responsiveness by digital companies and a basis for enforceable standards of conduct. Investigations into potential antitrust violations can illuminate and expose practices to political and consumer criticism.

America's founders shared with reformers in the 1890s concerns about concentration of power.[39] The Sherman Antitrust Act of 1890 reflected efforts to outlaw monopolies; subsequent legislative and executive efforts authorized legal challenges to concentrated power once held by Standard Oil and by AT&T. Senator John Sherman pushed for the initial U.S. antitrust law by arguing, "If we will not endure a king as a political power, we should not endure a king over the production, transportation and sale of any of the necessaries of life."[40] During the early twentieth century, Louis Brandeis and other antitrust reformers inspired aggressive antitrust enforcement, seeking individual liberty in economic as well as political and social life. Vigorous antitrust enforcement grew during the 1950s and 1960s in the United States, but then a hands-off approach toward economic concentration took hold in academic circles, and it continues to characterize executive action and judicial reasoning to this day.[41] Launched by University of Chicago law professor Aaron Director and advocated most effectively by his student Robert Bork, the Chicago School of antitrust law defines harms solely in reference to consumer welfare, which in turn is measured by short-term price effects on consumers. Deregulation of many industries, changes in capital markets, and corporate governance all contributed to waves of mergers in the 1980s and again in the 1990s, and to the tech conglomerates absorbing upstart rivals and other businesses.

Current reformers seek to revive the approach taken by Louis Brandeis and other antitrust reformers who argued for individual liberty in political and social as well as economic life. Those arguments are echoed a hundred years later by current academic reformers such as Lina Khan, Sabeel Rahman, Ganesh Sitaraman, Dina Srinivasan, and Tim Wu; media commentators; and Senator Elizabeth Warren.[42] Expressing concerns about alleged political bias in moderation, both Republican and Democratic political leaders in both state and federal governments pursue antitrust and other regulatory actions.[43]

The example of Facebook illustrates the argument for antitrust action. Facebook acquired ninety-two companies—including leading rival social media entities—between 2007 and 2019, and as of 2018, Facebook owned four of the top five social media services operating outside of China.[44] Its co-founder Chris Hughes explains that the social network is not a "natural" monopoly arising from the nature of the business, but rather grew through maneuvering that is or should be banned by law. Rival networks with constructive features could emerge if the anticompetitive practices of the biggest company were halted.

Yet recent U.S. antitrust policy has paid little attention to anticompetitive practices, economic concentration, market dominance, or effects of integration across distinct business lines, including control of infrastructure (and the information-gathering it enables) needed by ostensible rivals. Instead, antitrust enforcement pursues a narrow focus on keeping consumer prices from rising and averting risks of predatory pricing. This focus on consumer prices holds no relevance where the product is offered "for free." Yet "price" could be understood to include the data provided by consumers, often unknowingly, who leave their tracks by using platforms to communicate their preferences, interests, and connections. When companies exploit their power to obtain personal data and then use and sell it without candor to consumers, that looks like a serious problem that antitrust policy—or other regulatory policies—could and should address. Nonetheless, the lax attitude toward market power persists even when digital platforms undermine competition in the pricing and sale of goods by looking at how their goods-selling competitors use the platforms' infrastructure.

Tim Wu's critical work describing the rise of the "attention industry" and calling for antitrust enforcement paves the way.[45] Wu argues that "free products" offered by firms such as Facebook and Google compete in what is essentially a market for human attention, and they attract attention from individual users in order to resell it to advertisers for cash.[46] These practices may involve initially setting low prices for ads and then adopting higher prices that reflect market dominance—after a digital platform gains an advantage over its rivals.

Prevailing antitrust tools, focused on prices to consumers, miss the potential anticompetitive effects and potential harm to consumer welfare due to these practices. End users are bombarded by information, including much that is undesired. Beyond producing irritation and distraction for users from floods of ads, the business model relying on resale of end users' information to advertisers places priority on attracting attention over ensuring accuracy. Manipulating emotions is central to the strategy.[47] The strategy, in turn, can contribute to the rise of clickbait, fake news, revenge porn, and filter bubbles—risking harm to individuals and to the entire ecosystem of communications and information exchange.[48] Facebook's use of user data is at the core of its practice of choosing what news and other content people see; it highlights "Instant Articles" and other content in order to elicit "engagement," usually by seeking to trigger fear or anger.[49] This manipulation contributes to political and social polarization, affecting what people see, believe, trust, and distrust. For those troubled by these practices, a long-term consequence may be turning people away from media and then reducing people's ability to become and to remain involved in democratic self-governance.[50] And social media organizations may be deceptive in failing to disclose the ultimate uses of individuals' data. Facebook tracks and uses data about people who do not even have accounts.[51] Such treatment of consumer data should not be viewed as fair, much less fully consensual, because of Facebook's dominant position.[52] These issues, involving small players as well as large ones, have not for the most part presented the kind of harm to consumers contemplated by the U.S. Federal Trade Commission (FTC) or federal antitrust law enforcement, warranting a range of private and public responses. Still, the failure to restrict abusive data practices that are made possible by economic power should trigger governmental action.

A promising predicate for antitrust action would start by treating the appropriation and sale of user data as the "price" paid and examine the market power of digital platforms in that light. The editors of *The Economist* note that regulators and experts are increasingly seeing "personal data [as] the currency in which customers actually buy services."[53] The digital giants theoretically could be made to pay for users' data; because this is impractical, reforms should tackle the underlying misuse of power.[54] A 2019 German decision applies German antitrust law to find that the data gathering by Facebook exploits its dominant position and ties users to its network in ways that hurt competition.[55] Facebook contends that it faces direct competition from Snapchat, Twitter, and YouTube; while that is one way to look at those other services, they also could be seen as inadequate substitutes for Facebook, or even as complementary goods because users can employ multiple social media sites. Yet neither the FTC nor the European Union blocked the merger of Facebook and WhatsApp even as advocates warned of the negative impact on user privacy. Consumer protection or privacy laws, such as Europe's General Data Protection Regulation, would be more workable remedies than antitrust ones—or the prospect of antitrust action might prompt greater self-regulation or government protection of user privacy. The launch of landmark antitrust actions against the big platform companies may lead to very different policies in the future.[56]

Antitrust enforcement or reforms in its shadow could benefit the ecosystem of news, though only indirectly. Enforcement could take the form of orders forcing transparency about algorithms and other platform practices or requiring licensing of services or products to level the playing field for competitors. More competition for users of social media platforms could help produce better data collection, privacy, and innovation practices, with some ultimate impact on news reporting, quality, and distribution. Perhaps more competitive and consumer-respecting companies would generate more investigative reporting or other original newsgathering, or disrupt and possibly reverse the shift of advertising dollars from newspapers and other media to a few dominant digital platforms. There are a lot of ifs in this analysis, however. The scale of people's participation in existing social media and the digital platforms suggests that users are more satisfied than not, which falls far from the lodestar of antitrust law—increasing consumer benefit. The concentration of ownership may well reflect rather than cause defects in an industry. And antitrust law is at best an indirect and partial measure to tackle problematic data practices or the diversion of ad revenue from old media to new. In this light, those in charge of voting the stock proxies of retirement funds could do more to guard against further concentration of control and wealth in digital and media companies, or flex their power to seek some of the reforms to protect news and democratic values as well as investment returns.

Even if antitrust action is warranted, the remedy of breaking up the companies is not optimal for improving the gathering and distribution of news. The scale of and connections across a platform's services help secure the technical and financial resources for content moderation, privacy protection, and security. Concentrated investment in all these activities helps small communities both inside and outside the United States that otherwise do not have access to information. Not surprisingly, this is the position taken by Facebook chair and CEO Mark Zuckerberg as he advocates against government breakup of the company.[57] There is, however, a point here. The chief of Instagram—now a Facebook subsidiary—argues that breaking up its parent company could hurt efforts to battle bullying and harassment online because Facebook provides resources, reach, and expertise.[58] And even if large digital companies are dismantled, the resulting pieces would remain so enormous that the dangers from economic concentration remain.

Some note that the remedy of breaking up dominant U.S. tech companies neglects the value of their cybersecurity investments, which contribute to national security in the face of cyberattacks from China, Russia, and elsewhere.[59] A counterargument warns that large "tech companies are not competing with China so much as integrating with China, and their integration comes with threats to the United States."[60] Both views point to the need quite apart from antitrust considerations for public investments in nimble and continually inventive cybersecurity—and the need for the federal government to challenge and curb both large and small profit-hungry U.S. tech companies from entwining with other nations that pose security threats.

Most importantly, for the purposes of addressing the jeopardy in the world of news, an antitrust breakup remedy would do nothing to tackle the defects of reliance on private markets for the public good aspects of news media. As the concept is defined by economists, public goods offer value to the public but are goods that private markets are not well situated to provide. A street light, a road, and the military all offer benefits to the public but do not permit their creators to charge individual users a price; once in existence, a street light, a road, and the military offer benefits that do not diminish as more people use them, and it is not easy to exclude anyone who does not pay for them from using them.[61] When it comes to digital platforms, as long as a combination of advertising and subscription determines the revenues, and as long as competition for those revenues leads to heightened rather than lessened efforts to gain user attention and user data, unethical behavior can easily follow. Under current arrangements, there may not be a sustainable private market for continuous investigative reporting, especially in small communities, no matter how crucial such efforts are not only to people's interests in health, sports, and politics but also to the accountability of public and private institutions.

Multiple forces undermine the old business model for news production, but the dominant digital platforms are major reasons the legacy media are failing. The rise of a digital service—Craigslist—dissolved the revenue stream from classified ads that once supported local news. Technological advances allow readers to evade paywalls or share with screenshots and thereby undermine the effort of news media to recoup their costs. By unbundling sports, weather, food, and lifestyle stories from political, economic, and scientific news, the platform companies undermine a key aspect of the economic viability of traditional media. By making it possible for any user to post content, the platform companies bypass journalist gatekeepers. And the platform companies perfect the techniques of directing ads to users based on up-to-the-minute information about users' interests, gleaned from their traffic and aggregations of their purchases, searches, and demographic information. Even if digital advertising informed by data about individuals only modestly increases publisher revenues, it undermines older forms of advertising and the revenues that provided to broadcasters, cable companies, and newspapers.[62]

Some promise for improving the availability of varied news sources comes even within the limits of contemporary policy approaches to antitrust. One step toward producing diversity and competition would bring cross-ownership rules up to date, whether those rules are enforced by the FCC or through antitrust policies. Devising a safe harbor from antitrust concerns for newspapers or broadcasters that associate in order to survive would also offer a potential path for sustainability while acknowledging that their competitors increasingly are digital platforms.

Regulate Large Digital Platform Companies as Public Utilities

Theodore Roosevelt and Woodrow Wilson pursued regulatory innovations beyond antitrust laws to restore democratic control in the face of concentrated power held by private companies. Their notion of "public utilities" has new currency in the context of large digital platform companies. Government involvement is warranted when privately produced and owned goods or services are unusually important to and needed by society but economies of scale or other forces lead to corporate concentration.[63] These reasons apply to social media platforms and internet service providers. Public utility regulation recognizes that some resources are so essential to individuals and to society that the government is justified in issuing requirements around fair access and sustainability. Public utility regulation promotes goals that private companies do not normally have to consider. Regulation of a necessary good or service also helps guard against coercion that works by exploiting people's dependence, but it still permits private owners to operate for profit. Although competition exists for some parts of the news and media ecosystem, a few private companies with internet platforms exercise power over transmission and gatekeeping.

The concept of a public utility has roots in medieval European law; it was carried over to the American colonies, and for a time was expressed through charters offered to entities such as turnpike and canal companies.[64] The broad view of public utilities is not restricted to public goods that cannot be sufficiently provided through normal market incentives, but also includes goods important enough that private markets could yield corruption or abuse of those in need.[65] Justice Louis Brandeis maintained that dominant companies could be regulated more rigorously than other activities when they provide a necessity of life and when the forms of production pointed to domination by one or a few entities.[66] Water, electricity, and physical infrastructures for transportation and communications provided exemplary candidates for public utility status.

Treating major tech platforms as public utilities involves recognition of their control of debate, discussion, information sharing, and the distribution of news. That two-thirds of Americans get news on social media and millions of people rely on Facebook to share ideas and information gives a clue to how a few private companies are replacing the public squares, postal services, and other elements of public life that served as the backbone of communications. As legal scholar Sabeel Rahman explains, social media platforms operate now like railroads and twentieth-century telecommunications, the kinds of infrastructure that have justified public utility regulation.[67]

Through their scale and concentrated value, the major digital platforms have opportunities to combine different lines of business in ways that lock in users, close off competitors, and unfairly take advantage of their position. Their dominance yields profits that could be reallocated to support local news outlets and power that should be subject to public duties.

Social media expert danah boyd (who holds a research post at Microsoft) notes how Facebook has worked to become indispensable in the way that public utilities are—and public utilities "get regulated."[68] For example, Facebook has become a must-use platform for individuals whose schools or employers use the platform for internal communications. Google, similarly, dominates other sources for searching out information and uses its proprietary algorithms for continual modifications of its services based on the data it gathers about its users.

More concretely, cities and other governments are contracting with Google and other tech companies to provide public services and communications involved in security, data, and management of other public utilities.[69] Professor Julie Cohen shows how internet platforms are not merely data collectors nor even simply networks; they are "infrastructures" that facilitate production of other goods, just as roads and electric power grids do.[70] Because the internet platforms serve consumers, sellers, and advertisers, the companies can reduce prices to one group while charging fees to others—and use their scale to create enduring competitive advantages over other players in markets. This is the kind of price discrimination addressed by public utility regulation.

Public utility regulation of digital platforms would focus on ensuring fair access and treatment, protecting the privacy and safety of users, and establishing transparency and forms of accountability. Guarding against manipulation, a goal more difficult to achieve, could also be explored. Such goals could be pursued through a variety of means: oversight by regulatory bodies, self-regulation reported by companies, professional standards monitored by civil society associations and watchdog groups, public options offered to compete with private ones, required access to coordinated data, programming tools to permit monitoring by the government or third parties, and other approaches.

Classifying the internet itself or some elements of it as a public utility is one predicate for net neutrality (federal or state regulation preventing internet service providers from blocking or slowing transmission of some content in preference to others; see Chapters 2 and 3).[71] Even as the 2015 effort to ensure access to broadband net neutrality survived judicial review under a public utility theory, it fell to the change in political administration that took place after the 2016 presidential election.[72] Public utility regulation of Google, Facebook, or Amazon would be a new application of the concept, one that arguably accords with the focus on the risk of consumer misuse.[73] As sources of communication and information, the platforms do differ and do have competition, but their position allows them to take and use personal data, without disclosure, in ways that can be characterized as oppressive. Social media and other services provided by Google, Twitter, and Facebook are more and more treated like a necessity. Public schools and private organizations depend on them for communicating with their communities. Individuals depend on them for information and connections with others.

Regulating large internet platforms as public utilities could take place through a commission made up of public servants. Regulators would oversee market entry, exit, and expansions; standards and terms of service; and disputes and complaints by users. Oversight could require annual reports, guard against unfairness in service provision, and ensure due process and transparency while leaving the platforms as private entities still able to restrict hate speech, clickbait articles, or inflammatory or untrustworthy news. Some propose protections against exploitation and subordination of users by separating platform functions in situations where combined activities create incentives for self-dealing or other conflicts of interest; others suggest articulating and enforcing public obligations of fairness and service to historically underserved individuals and addressing harms caused by misleading information. Regulation tailored to large tech platforms would require data sharing in order to enable oversight by public or nonprofit watchdogs, or even to enable start-up rivals to enter the field. Enforcement duties could be conducted by existing governmental entities such as the Federal Trade Commission, the Department of Justice, state attorneys general, or a new oversight commission, girded against risks of regulatory capture by industry.[74] State agencies, in the absence of federal law, could serve this function, but the risk of inconsistent rules across multiple states and jurisdictions in the absence of a physical presence in the state would make compliance difficult. What is needed are practices and institutions to monitor and challenge the exploitation of consumer data as well as unfairness and transparency in the conduct of digital media. Just as the civil service, and independent bodies arose to check other powerful actors, new checks and balances are needed to address the intentional and unintentional negative effects of the large internet companies.[75] The British government in this spirit has created a new regulatory entity, the Digital Markets Unit, to protect consumer data and to assist smaller news outlets from the domination by large digital platform companies.

At a minimum, the public utility framework would allow government enforcement of self-regulation and content moderation by the tech platforms. The government could require digital companies to report how they are providing security against flooding users with unwanted messages and harassment; how they are ensuring fair access, fair data acquisition, and transparent data use; how they are investing in enforcing the digital platforms' own rules; and how they are upgrading the fact-checking or credibility assessment tools they make available to users. Or the government could require that the tech companies contribute financially to independent entities that rate the reliability of internet news posts. Currently, digital platform companies do not adequately enforce their own internal terms of service.[76] Google AdSense, for example, seeks added value and supposedly does not take payment for sites that simply copy and paste content from other sites—yet that is precisely what some sites that sprang up around the 2016 election did, and those sites were still able to make money by having AdSense place ads on them.[77] Seeking to provide transparency, Google itself reports on the amount spent on ads during the 2020 election campaigns, it also, as of 2021, labels ads that violate Google's own policies—but does not release information about which policies have been violated. Alerting users to manipulation by ads that are disguised as content that is not subsidized would be another potential area for improved and serious enforcement.[78] When Microsoft gives users of its browser free access to rating labels provided by a private, independent fact-checker, it is undertaking a duty that government could enforce.[79] Even if newly articulated duties of care to users are adopted and enforced, however, self-regulation and content moderation by the tech platforms would leave in place many of the incentives that produce the current problems with news infrastructure.[80] The conflict of interest between profits from controversial content and the task of moderation is too profound to leave to self-regulation.[81] Setting standards of care—including treating the platforms as "information fiduciaries"—would be compatible with public utility regulation.[82]

New Fairness and Awareness Doctrines

When reliance on the broadcast spectrum made communication opportunities scarce, the FCC required radio and television licensees to provide balanced, fair treatment of controversial issues. Although cable and internet options have undone that particular scarcity rationale, the new scarcity is the attention of viewers and readers. This scarcity of attention—and also the public interest in guarding against demonstrable risks of users' confusion and exploitation—could come to justify a new fairness doctrine. It would apply to digital platforms such as Facebook and Google. This would require new thinking; it could build upon the First Amendment right to receive information. A new fairness doctrine could enshrine journalistic norms of public service, curiosity, and responsible presentation of competing views while working to minimize risks of government censorship. And, perhaps less controversially, a new "awareness" doctrine would improve users' knowledge of the sources and nature of what they receive and also the patterns of their own engagement.[83] An awareness doctrine could involve content distributors in devising labels to distinguish news reports from opinions or unverified claims—without enshrining false equivalencies between factual and demonstrably false materials; it could provide to individual users upon request reports of their patterns of materials consumed and produced—patterns the companies already track. Challenged to improve their users' discernment of value in content they read and they post, platform companies could compete to develop varied methods.

Whether enhancing users' awareness or promoting fairness in curating material, the platforms would not become censors; instead, they would search for and share competing views. Google News has already announced an aim to connect users with a "broad array of perspectives to help [users] discover [their] own informed opinions" and to find "trustworthy" information for users through publishing partners.[84] Still being piloted is a Facebook News App intended to highlight stories that are factual, diverse, original, timely, fair, and local.[85] In addition to government enforcement of voluntary promises by platform companies to provide services, how about regulations requiring those companies to make such services available? Justifications for such regulations that should survive a First Amendment challenge turn on the limits of human attention, on the vulnerability of each user to being overwhelmed by floods of digital materials, and on the crucial dependence of democratic self-governance on access to meaningful information.

Professor Cass Sunstein, who considered and later rejected the idea of a fairness doctrine for the internet, urges a "serendipity" element through which Facebook and similar platforms would enable people to encounter material quite different from what they usually absorb or prefer.[86] Algorithms are already used to narrow what people receive; they could be modified to expand what people receive. Government requirements would stimulate innovation to help people see material that broadens rather than narrows the views and inputs they encounter, and feedback by users and observers would help develop techniques for breaking out of the filter bubbles and echo chambers crafted by current media and digital companies.[87] Individuals could have enhanced choices to see (or not see) a broader array of content than what their own history would generate; the question is whether the default setting—unknown to the vast majority of users—narrows or broadens what they see. The risk that a heavy-handed government might suppress certain messages can be considerably reduced by putting the burden of creating the methods for finding and sharing competing views on the shoulders of the digital platform managers and by ensuring options for users. The tech companies could encourage users and critics to give feedback and could compete on methods for offering access to an expanded range of views. For example, in 2017 four undergraduate students at the University of Chicago created FlipSide, an artificial intelligence platform that uses an algorithm to assess political ideology and then provide users with news stories and opinion articles from opposite points of view.[88] The government could direct internet platforms to devise a rating system to distinguish news, analysis, and opinion, much as the Telecommunications Act of 1996 directed distributors of video programming to establish voluntarily rules for rating sexual, violent, or indecent material.[89] There are good reasons to keep government away from any editorial or censoring powers, but government can avoid those roles while still requiring digital platforms to deliver ways to provide readers with contrasting views. At a minimum, government can require platform companies to develop labels or ratings of content to distinguish news reports, analyses, and opinions and try to prompt a culture expecting the use of such labels. More ambitiously, the law could require the internet platform companies to give users options for receiving information that diverges in point of view from their habitual sites; a middle position would require platforms to give users the option to receive for their personal use data to compare their habitual sites and news sources with patterns that others see.[90] Policies inviting companies to improve their users' awareness or to ensure easy access to contrasting viewpoints would revitalize the recognition that it is the rights of those receiving communications to hear varied sides of controversial questions, not just the rights of the owner or speaker.[91]

Nothing here is meant as an endorsement of the proposal introduced in 2019 by Senator Josh Hawley, Republican of Missouri, to terminate the immunity from suits regarding third-party content that social media and platform companies receive under Section 230 unless they receive certification from the Federal Trade Commission of their political "neutrality."[92] Government certification of "neutrality" is both impossible and undesirable in a nation devoted to constitutionally protected communication and expression. Proposals like Senator Hawley's would put the federal government in the impossible (and illegal) position of deciding whether a company is politically neutral, and they would do nothing to enhance the expression of competing and contrasting views.[93] Such proposals would also bring the government into decisions about content, triggering the most stringent form of judicial review under the First Amendment.

VITALIZE PROTECTIONS AGAINST HARM AND ABUSE

A second strategy for reforms—bolstered by public utility and antitrust legal authority and long traditions of contract, tort, and consumer protection laws—involves vitalizing public and private protections against deception, fraud, and manipulation. Strengthening capacities for law enforcement means strengthening both norms and implementation. Also valuable would be steps to equip civil society groups to monitor abuses and to press for better media and internet practices, because most individuals lack the time and the capacity to do so. Private leaders—including corporate leaders—could do more to hold social media companies responsible for egregious misconduct by choosing where to direct their ad dollars.[94] Laws governing commercial relationships and duties of care, if bolstered and enforced, could strengthen the news media ecosystem.

Enforce Contractual Terms-of-Service Agreements

Terms-of-service agreements are enforceable contracts, but few people actually read the fine print of such agreements, especially when the mere click of a digital button—a typical necessary step to using an online service—is treated as acceptance of those terms of service.[95] Experiments and survey research confirm that people do not pay attention to online terms of service agreements; few even read printed contracts. Changes in website design may prompt people to read the terms of service, but they would still most likely lack sufficient time and knowledge to adequately understand them.

And people lack leverage to resist such agreements if they want access to Google, Facebook, and other resources—and certainly lack power to negotiate for changes to such terms. But such agreements are enforceable.[96]

Legislatures can require consequences for violating terms-of-service agreements, as Congress has done with the 1986 Computer Fraud and Abuse Act. This law establishes a federal crime of accessing a computer without authorization, and such authorization includes elements within terms-of-service agreements.[97] Acknowledging that companies could simply exclude anything meaningful from their terms of service, legislative or consumer protection administrative bodies should establish codes of conduct for companies or require the companies to establish them—and create legal consequences for failures.

One reform option would be to enact legislation creating fines or other consequences for violating agreements promising reliable service. Another would be to require—and then enforce—promises to guard against fraud and fakery. Even with First Amendment limits on compelled speech—expression required by the government—disclosures can be required where commercial speech in question is misleading or unlawful.[98]

Why should users lack the ability to get redress for the failures of internet platform companies to adhere to their own guidelines and to protect users as they promise? Consider the scandal of Cambridge Analytica, a private consulting firm that, beginning in 2014, obtained the personal profiles of some fifty million Facebook users and data about their friends through a researcher working in violation of the agreed arrangement with Facebook.[99] Then the data were allegedly used to target voters in President Trump's campaign for the presidency. These mistakes suggest that there should be some legal protections for individuals that companies would not be able to circumvent through click-through agreements. Nothing in the First

Amendment should bar holding internet companies responsible for the promises they make.

More difficult to regulate, consistent with First Amendment concerns, are sites producing a large amount of false material. Giving the government the job of removing misinformation and hateful comments would be giving it too much authority to suppress speech. But the government could require private companies—if treated as public utilities—to report how well or how poorly they enforce their own rules governing their terms of service. Despite emerging First Amendment challenges to government-mandated disclosures on the grounds that such disclosures constitute compelled speech, powerful rationales for some disclosures can outweigh burdens, at least where the expression in question is of a commercial rather than political nature.[100] For example, damage to electoral integrity and personal privacy could justify requiring disclosure of the country of origin of the posted material—and whether it is generated by a person or a "bot."[101] Such information need not breach anonymity of the speech, which the Supreme Court has deemed protected by the First Amendment (yet even protection for anonymity could be reconsidered under circumstances of grave harm or pragmatic assessments about the actual risks of deterring speech).[102] Disclosures about where messages come from help listeners test the reliability and meaning of what they encounter. Just as a state may forbid a person from concealing their face in public or on private property without the owners' permission, digital platforms could require disclosure of the identity of those behind particular speech or to restrict the distribution of posts without such disclosures. The values of security and the integrity of elections can at times outweigh the arguments for anonymity as a way to guard against harassment, abuse, or privacy.[103]

Similarly, disclosures making critical information transparent would enable independent auditors—whether based in government, academia, or nonprofit organizations—to hold internet companies accountable. Companies already disclose some information but do not do so clearly and consistently. That means it is difficult for anyone to know how the companies comply with their own rules, precisely what user information they collect or share, and what choices go into their algorithms. Concerns about compelling speech through mandated disclosures must be weighed against the harms permitted by dominant tech platforms. Rules against defamation, intentional infliction of emotional distress, consumer fraud and perjury, and false claims about disasters and terrorist attacks are permitted within the First Amendment, and so are mandated disclosures under some circumstances where consumers, investors, watchdog groups, and voters need information to protect their interests.[104]

Regulate and Enforce Fraud Protections

Traditional legal tools meant to protect individuals from misrepresentation, fraud, and violation of contractual terms operate at both state and federal levels. Users of social media and other online sources could complain about misrepresentation when material pushed to them conceals its source, presents false or harmful information, or violates terms-of-service agreements. Individuals, advocacy organizations, and government actors such as state attorneys general and the Federal Trade Commission can investigate and pursue enforcement of common law and statutory protections, but reforms of both law and practice would be needed to make these tools effective in the context of digital news and advertising providers and platforms. Although it has no authority to fine abusers, the Federal Communications Commission asserts its authority to investigate complaints about intentional distortion of broadcast news "if there is documented evidence of such behavior from persons with direct personal knowledge."[105]

State consumer laws can be a resource for experimentation, although national and even global regulation is needed in the long term to deal with global companies. In 2011, a state court interpreted California's Consumer Legal Remedies Act to permit claims against a defendant for use of "false news, stories and endorsements" that might lead a consumer to believe incorrect information about the defendant's product.[106] In that case, the defendant operated an online auction site, reaching users through sponsored advertisements and fake news stories. Embedded marketing blurs the line between advertising and news, straining the rationales for distinguishing between less-valued commercial speech and more-protected types of speech. Consumer interests in understanding how their news is found and delivered could justify compelled disclosure of the choices made by platforms in their moderation and algorithms without jeopardizing claims about proprietary information.[107]

The European Union has at times effectively forced Facebook and other digital companies to remove fake accounts. The United States could and should impose and enforce requirements to remove fraudulent accounts from digital platforms and to be transparent about their efforts. In the United States, more difficulties arise around regulation both because of First Amendment concerns and because the country hosts so many sites producing a large amount of false material.[108]

Nonetheless, even though the Supreme Court has interpreted the First Amendment to provide some breathing room for lies,[109] the First Amendment has from the start coexisted with laws against defamation and fraud.[110] Ongoing government enforcement of truth-in-advertising and corporate disclosures has not contravened the First Amendment, nor should it in the future.[111] Social media platform companies already have voluntary agreements with the governments of many countries to root out fraud, and those agreements should be enforced.[112]

The Federal Trade Commission prohibits "unfair or deceptive acts or practices in or affecting commerce." The FTC successfully challenged fake "news reports" claiming to be objective assessments of a weight loss product; the reviewing court concluded that the "fake news website likely would mislead a reasonable customer.[113] Liability can even attach to those who approve or contribute to the creation of the fake news sites, according to one court addressing fake news sites marketing bogus weight loss products.[114] Yet platforms enabling the circulation of fake news are immunized by law even when they carry misleading ads and stories produced by third parties unless the platform itself directly participates in the production of the problematic material and plans to share it—unless, as discussed earlier, that Section 230 immunity itself is changed. Even though false speech may receive some constitutional protection, responsibilities for purveying falsehoods can attach to private actors who select and moderate content. And even given protections for anonymous speech and limits on compelled speech, the constitutional analysis should be nuanced enough to permit required disclosure of foreign funding for political ads.

Require Transparency About Choice Architecture and Curation

In the print and broadcast worlds, by conventions and by law, readers, listeners, and viewers are informed about who is the author, editor, or other source of news and other communication. The U.S. Federal Communications Commission prohibits hoaxes and intentional "rigging or slanting" of the news.[115] Even when a specific author is identified as anonymous, the editors and publishers are identified and are liable for claims of defamation, copyright infringement, and other harms. The digital design of the internet and the legal rules around it have both deprived users of transparency about sources and rendered platform companies not liable for harms arising from material offered by third parties using an internet site (with a few specific exceptions). Moreover, with devices such as "infinite scroll," "autoplay," and "eliminated natural stopping points," social media platforms push content to people that they have not selected—and do so to promote prolonged engagement by and data gathering about users.[116]

Laws could require platform companies to post warnings that material may not be reliable and regularly explain their curation practices. Popular sites deploy psychological expertise and massive amounts of data about individual users' behaviors to "nudge" people to click on and share false information.[117] Using nudges that are not transparent can be viewed as the kind of exploitation that justifies state intervention. Cass Sunstein, a leading advocate of nudges to shape behavior, emphasizes the importance of transparency to check exploitation and unethical abuses.[118] These nudging practices, though, are designed particularly to operate without people's awareness.[119] They raise concerns about manipulation, which is particularly concerning because it is hidden.[120] The covert use of online tools for "hypernudging," based on highly personalized information and utilizing intrusive design, harms individuals' autonomy and threatens the development of preferences and deliberation that undergird collective self-government.

A reform worth considering would require tech companies to disclose their practices in more detail than requesting the user to click and thereby give general consent to whatever the service does. Determining what degree of data collection and nudging is acceptable would require deliberation and expertise. There are competing goals at work, however; the reform would cause some people to change their behaviors by selecting anonymity and encryption, which would make moderation by the platforms more difficult (unless there are changes in the treatment of anonymous participation). Desirable reforms would keep both privacy and effective moderation in mind, and also spur innovation around personal data aggregation and financial models other than "surveillance capitalism," markets that rely on gathering and selling users' personal data aggregated across the internet.[121] Some digital companies pursue subscription fees rather than advertising; although this is not an option for everyone, a digital platform and information aggregator relying on subscription fees could avoid coaxing users to give up their personal data and might increase competition for quality content. A related idea would impose on a large platform company the requirement of creating distinct but interoperable versions of its services with varied financing options (ad-supported, free in exchange for personal data, or money charges) as a remedy for antitrust or public utility concerns.[122] Making sure that internet companies bear more of the costs from their use of personal data could result in a push for alternative revenue sources for news and information besides targeted ads—and perhaps improve variety and increase the number of higher-quality news sources.

Support Civil Society Efforts to Monitor and Protect Individual Internet Users

Most individuals lack the time and expertise to ensure that internet companies adhere to their commitments or comply with regulations. A typical American would have to spend 250 hours a year just to read the terms-of-service agreements they have accepted.[123] So supporting the work of nonprofit and governmental consumer protection efforts is a crucial part of protecting individual internet users—and, in turn, reducing the practices that impair responsible journalism and foster the spread of fake news, conspiracy theories, and hateful messages. The tech companies have information that should be available to others to enable necessary accountability measures.

In part, it will take tech tools to break out of the problems that tech tools have created. Tom Wheeler, former chair of the FCC, and Wael Ghonim, a former Google employee who helped spark protests during the Arab Spring, propose requiring social media platforms to coordinate through "open application programming interfaces"—not so that one platform can steal another's secret social media algorithms, but to enable third parties to build software that can monitor the consequences of those algorithms. Wheeler argues, "The best approach is to share information and ideas to increase our collective knowledge, with the full weight of government and law enforcement leading the charge against threats to our democracy."[124] Most individuals will have no ability to make sense of the materials even if they are disclosed, but advocacy organizations, academic experts, and government regulators could take up the work of analyzing and explaining the choices that affect what people see.[125]

What is needed is not mere disclosure but actual access through interfaces, allowing real analysis. Interfaces already allow Google Maps to work with Uber. An interface can protect the privacy of users as well as the secrets of algorithms even as it makes it possible for others to track who purchases social media ads, the extent to which those ads are accelerated and distributed, what content is deleted, and how much of that content is spread before deletion.[126] To comprehend the computational and analyzing power of private networks, government and public interest groups need to be able to see what so far has been hidden from view.[127] Requiring more transparency would allow people beyond the platform programmers to understand and critique the curation of content received by users. The United States would need not only to adopt requirements in this direction but also to persuade the European Union to adjust its privacy requirements to permit at least enough transparency for educational and nonprofit researchers to serve as watchdogs. Governing boards of private digital platform companies and telecommunications companies should establish risk committees that obtain and review audits of how their systems work and how they deceive. Governments can promote or require such efforts. The boards and chief executives of media companies should make it a priority to improve the credibility of the content they distribute.

AMPLIFY AND SUSTAIN NONPROFIT AND VARIED SOURCES OF NEWS AND ACCOUNTABILITY

Public resources to support journalism and news have been a feature of American life since shortly after the founding of the nation. Early postal subsidies permitted newspapers to be sent through the mail at reduced rates (which did not fully cover the costs of distribution) and encouraged the free exchange of newspapers and periodicals among their producers.[128] Taxpayer dollars also support public broadcasting, although not to the degree found in the United Kingdom, Japan, and other democratic nations.[129] Several European and Nordic nations also employ tax exemptions for nonprofit journalism and direct aid to for-profit journalism in order to promote quality print (and, in some instances, digital) media.[130] With the challenges facing newsgathering and distribution, especially in local communities, public support for public and nonprofit media should no longer be treated as optional and instead must be secured and expanded. Public and nonprofit media options—aided by direct and indirect public support—can fill news deserts and other gaps left by profit-oriented companies. Public and nonprofit media also provide crucial competition and can stimulate for-profits to win viewers by doing better.[131] The operational press that constitutional freedoms of expression and the press support is a predicate for democratic participation and governmental accountability.

Support Nonprofit News Sources with Tax Exemptions, Deductions, and Credits

A tried-and-true method of public support that also advances freedom and diversity of views is tax exemption for nonprofit organizations. One congressional proposal would create a tax credit for local news organizations making new hires.[132] Exempting all news organizations from taxation may not be workable, but preserving and allowing tax benefits for media organizations organized as nonprofits is an essential avenue, as the for-profit model for the press falters in the face of competition from internet platform companies. Promising nonprofit efforts include ProPublica, the *Texas Tribune*, and the *Salt Lake City Tribune*.[133] Each works to create multiple revenue streams, including grants, donations, and subscriptions; each makes use of the public subsidy accorded through nonprofit status.

In the decade since a 2009 call for greater philanthropic support for journalism, private contributions have quadrupled, and nonprofit newsrooms receive about 40 percent of their income stream from private foundations.[134] Local news initiatives are among those attracting philanthropic support.[135] Government should preserve tax deductions for such contributions.[136] Investments in nonprofit media not only provide independent sources of news driven not by profits but by professional norms but also encourage competition, and an ethos of sharing their reporting can work to improve the larger ecosystem of news and media.[137]

Federal and state tax powers can be used to burden enterprises that impose costs on society and to reward those that bring benefits. These chief options are to impose new taxes on for-profit companies and to strengthen or deepen tax incentives for philanthropic and nonprofit activities that advance the common good. For example, federal taxes could be imposed to deter or limit "surveillance advertising," which targets users by integrating their stated interests with their past behavior and other personal data. Economist Paul Romer's proposal along these lines intends to discourage surveillance advertising and encourage alternatives such as subscription-based business models.[138] Media policy researcher Ethan Zuckerman proposes taxing advertising that not only tracks users but also integrates masses of personalized information, with the tax revenues used to support public service digital media.[139] Others suggest simply taxing the platform companies to fund journalism, much like proposals for a carbon tax to finance environmental protection and climate change mitigation.[140] Tax credits intended to promote new hiring of journalists or conversion of for-profit media to nonprofit could also bolster newsgathering and reporting.[141]

Tax rules can spur private donations to nonprofit organizations in two ways: tax deductions or credits encourage donations, while tax exemptions offer relief to nonprofit organizations from some costs. Philanthropy provides some support for nonprofit news operations, including the big-data analyses by ProPublica and media watchdogs that enable journalists and citizens to track the local dimensions of national events.[142] None of these proposals introduce censorship or constraints on particular views; none abridge the freedom of speech or the freedom of the press. All build on steps taken in the past. These proposals and others like them could revitalize the news that democracy needs.

Fulfill Public Obligations to Public Media and Media Education

Telecommunications intended to serve the public seek excellence, diversity, accountability, independence, and innovation in informing, educating, and entertaining those within its reach.[143] Despite claims that PBS and NPR are unnecessary or obsolete,[144] three-quarters of Americans polled favor maintaining or expanding governmental support for public broadcasting.[145] There is a crucial need for both an "information commons" and for the competition that public service communications have been able to provide to private media. Surveys rate public media as the most trustworthy news organizations now and for the past fourteen years.[146] Public support for media can include ownership, subsidy, or other aid, and must always be provided with attention to include rather than suppress diverse views.

Public media makes a difference in addressing holes such as local news deserts created by private actors.[147] It could create digital platforms for community news and information, including a capacity to collect and analyze big data in the service of cities and towns.[148] Given more resources, public media could expand investigative journalism, the creation of documentary films, and exchanges of community news and information. Public media often pursue background and multiple explanations for events in ways that commercial media do not.[149] Granting public media the flexibility to generate funding through underwriting by private sources would require policy changes; so does adjusting the Copyright Act's treatment of public media to reflect new distribution platforms.

The commitment to and stability of publicly supported media has been stronger in countries other than the United States. By locating responsibility for broadcasting content creation in the post office, Great Britain funded the British Broadcasting Company with license fees first on radios and then on televisions while shielding content choices from government control over programming.[150] The resulting quality of programming has earned both trust and influence in other European countries pursuing public service media.

In the United States, commercial development dominated, but an early chair of the Federal Communications Commission, Newton Minow, pressed successfully for critical responses to the market's failures and for the creation of an infrastructure for public service media.[151] The All-Channel Receiver Act of 1961 required that new television sets be made to include ultrahigh frequency

(UHF) channels as well as very high frequency (VHF) channels, and this made room for public service media even where no licenses for VHF channels were available. The Communications Satellite Act of 1962 provided wireless capabilities, which local public broadcasters in the United States have used to distribute their programs across the country. Eventually, Congress strengthened public broadcasting with the Corporation for Public Broadcasting, supporting radio and television offering children's programming, documentaries, arts and humanities shows, and other content that commercial broadcasters had failed to offer. Sometimes, such as in the case of children's educational broadcasting—notably *Sesame Street*—the public content attracted strong enough followings to inspire similar efforts by commercial broadcasters.[152] In the midst of political polarization, NPR and PBS remain among the most trusted media.[153] States can and should commit to public support for regional and local public media.[154] Even though a recent report has shown that there is less trust globally in public media than in the past, a majority of people still prefer news without a particular viewpoint over partisan outlets.[155] Public support for broadband and other media infrastructure is also critical.[156] Federal, state, and local funding should help ensure local news coverage through print, broadcasting, and digital means.[157] The educational and public service dimensions of public media are essential, as efforts during the COVID-19 pandemic demonstrated.[158]

Funding for public media in the United States has remained fragile. Unlike Britain's public option, financed through the annual television license fee, public media in the United States competes with all other priorities through the political appropriation process. Efforts to secure public broadcasting finances through a 5 percent tax on television and radio sets failed from the start because American unions viewed such a tax as regressive, imposing relatively more on the poor than on the wealthy.[159] The government could set aside for public media a portion of the proceeds from frequency auctions, or a portion of the application and renewal fee for broadcast licenses; alternatively, it could reduce the tax write-off for advertising. Mechanisms to ensure sufficient insulation from government actors and actual content decisions already exist and can be bolstered; new models that engage and inform readers who do may not otherwise seek out reliable news sources can also be designed.[160]

Some argue also for a publicly funded alternative to Facebook and Google.[161] Congress could even condition funding of such a platform on a design that promotes deliberation rather than relying on subscription fees polarization, solutions rather than shouting matches.[162] Reliance solely on governmental support for a public platform creates vulnerability to political trends or capture by particular interests.[163] Public-serving media needs to be able to rely on multiple funding sources, and it also needs mechanisms to insulate the support it receives from public sources from the whims of particular political moments.[164] In another model of content-neutral aid, government could match funds raised by nonprofit newsrooms from their own communities.[165]

Above all, public support should be substantial, stable, and secure. In their book *The Death and Life of American Journalism*, Robert McChesney and John Nichols calculated that the level of government subsidy given to the American press in the 1840s was the equivalent of $30 billion in 2010 dollars, which is far more than the actual support the government provided in 2010.[166] Benchmarking based on the past, on efforts in other nations, or on estimates of needs and benefits would lead to considerably higher investments in public media than are current in the United States.

Media education, equipping people to become informed and aware consumers of media of all sorts, should highlight digital media practices and risks.[167] A 2012 study shows that digital media literacy is associated with greater political engagement and with exposure to diverse viewpoints.[168] Media literacy education points students to what makes sources trustworthy, how to separate fact from fiction, and how to think critically about information and about the choices behind stories that are told, produced, and distributed.[169] Critical thinking for assessing news fundamentally means knowing to check for balance, adherence to facts, citation of sources or explanation why the sources are anonymous, responses by anyone accused of negative conduct, and skepticism about claims of conspiracy or secret knowledge.[170] Highlighting efforts by providers to distinguish stories exchanged by friends from content produced or vetted by professionals would assist educational efforts aimed at helping users distinguish different materials generated and posted in different ways.[171] So would labels identifying how asserted "facts" in posts have or have not been verified.

Each of the approaches—hold internet platform companies responsible for their conduct and its effects on the availability of news, protect readers and users from abuses, and ensure the availability of independent nonprofit and public media and media literacy education—has limitations. No one approach can fix the problems giving rise to news deserts, "fake news" circulating on social media, and the loss of viable business models for legacy media, and no one of them is required by the Constitution. Nonetheless, they are not barred by the First Amendment, and they are inflected by its guarantee of freedom of expression and its presumption of a free press. In combination, they could turn matters around.

CHAPTER FIVE

. . .

Coda

SELF-GOVERNMENT AND CHECKS on public and private power cannot work if people do not have access to independent information. Basic health and safety also rely on timely, reliable information. The guarantee of free speech and a free press in the U.S. Constitution's much-exalted First Amendment presupposes the existence of an independent press. That predicate is in jeopardy due to the disruptive effects of internet platforms taking the advertising revenue and sometimes the content of legacy media without investing in the production of news. It is in jeopardy in many local communities because of consolidations stripping local news operations of staff. It is in jeopardy because of mistakes made by long-standing media operations, mistakes including tardy accommodation to the demands and challenges of digital distribution and readers becoming accustomed to the "free content" model of the internet platforms. And it is in danger as the confluence of social media platform "nudges," decimation of legacy media, and political assaults pulverize confidence in truth and shared understandings of something as basic as election returns. This is not a time, though, for laying blame; it is a time for responsible action.

An aggressive libertarian reading of the First Amendment, growing in the courts, may suggest that government's hands are tied. That reading is wrong. It is belied by deep and extensive government involvement in funding, shaping, and regulating media and the circulation of information. From the early post office subsidies and government investments in and regulation of telegraph and radio to antitrust enforcement shaping broadcasting and cable, and on to government financing of research producing the internet, legislated immunity from liability for platform companies unavailable to legacy media companies, and the creation and shifting tides of public media, government directly and powerfully influences media and information access—indeed, the entire ecosystem of news. The "press" does not just exist, unchanged by governmental, economic, and technological developments; policies in each realm profoundly shape the press. And constitutional protections apply not only to guard against government abridgment but also advance the rights of access to the information that people need to be informed as voters and participants in their communities. The First Amendment thus has an affirmative face as well as a negative injunction.

The Supreme Court has so acknowledged, as do thoughtful scholarly understandings of the relationship between freedom of speech and democracy. Relying solely on for-profit companies does not secure credible news sources: a hands-off government instead has helped to create companies that permit manipulation, division, distrust, and tampering of election-time information.[1] Even the platform companies acknowledge that the issues they face exceed their competence and abilities. Restoring trust, protecting against hijacking by foreign powers, cultivating informed and skeptical citizens, and strengthening the ecosystem of news and democratic governance requires public action. With the entire project of democracy in danger, federal, state, and local governments can and indeed should be obliged to act—while remaining as neutral as possible toward content and viewpoint in private speech. If judicial readings of the First Amendment prevent such actions, the courts would be turning the Constitution into a suicide pact.

That is both an undesirable and unnecessary future. The First Amendment poses no barrier to ensuring the same laws applicable to other carriers and producers of communications to internet platforms. Internet platform companies should be treated as responsible actors. Government (and private news producers) should require payment from the tech companies for news they circulate or allow others to circulate on their sites. Internet platforms should be subject to the same liabilities as any distributor, and perhaps as any publisher, or else immunity should be conditioned upon their embrace of and compliance with transparent rules of moderation. Tech platform companies' use of individuals' data should be taxed. The internet companies should also be investigated for antitrust violations and regulated accordingly. Treating the large digital platform companies as public utilities would permit regulation, including potentially a new fairness doctrine requiring responsible exercise of the platforms' moderating capabilities or an awareness doctrine assisting users in navigating both the content and their own uses of it. The platform companies can and should assist users in distinguishing opinions from news reports and in comparing their own exposure to particular sites with what others see. Government should help protect individuals from harm and abuse caused by the large internet platforms by enforcing terms of service and codes of conduct, regulating and enforcing protections against fraud, requiring transparency about how those platforms curate content and nudge users, and supporting civil society efforts to monitor their behavior. Attention to the depletion of local news in particular is vital and calls for efforts across public and private sectors.

Government should amplify and sustain varied sources of news and accountability through nonprofit and public media. Tax policies, including the use of exemptions, deductions, and credits for nonprofit organizations, are a promising tool and could even be extended to any newsgathering and news-producing company. Taxing the huge tech companies and investing in public and private systems of auditing and assessment are tools within the constitutional authority of the federal government. And stable and meaningful support for public media and media education should be treated as government obligations.

Although none of these initiatives alone will fix the crisis in reliable news locally and nationally, efforts on each dimension, if combined, would help. The First Amendment is not a barrier but instead a basis for these actions. Alexander Meiklejohn introduced valuable thinking about the meaning of the First Amendment entirely in terms of democracy.[2] Those ideas have not been fully adopted by the courts, yet they are timely and worth attention at a time when democracy itself seems fragile and the news industry is faltering. Meiklejohn also had a healthy ability to look forward, not just backward, and he emphasized how much we need to remain open to change. He wrote, "We must accept and applaud the assertion that the Constitution is an experiment, in the sense in which all life is an experiment."[3] The success of that experiment depends on our ability now to enable in our rapidly changing world the production, distribution, rigor, and trust in news that are essential to a democratic society. The framers of the First Amendment understood this when they committed to protect the freedom of the press—the only private business named in the Constitution. The tools to do so are within reach.

Acknowledgments

. . .

Writing at a time when democracy, news industries, trust in fundamental public and private institutions, and faith in the very enterprise of truth-seeking are imperiled is a daunting challenge. I find hope in the ongoing efforts by so many young people working for renewal and revision of these foundational commitments, and in the generosity and interest of so many scholars, journalists, friends, and strangers.

Journalism drew me in when I was young. Print and broadcast media brought me the civil rights, women's, and environmental movements, the Vietnam War protests, assassinations of revered leaders, and exposes of corruption and I became a student journalist. During Watergate, the fall of the Berlin Wall, and other memorable events, I admired the critical role of journalists, and I remain grateful to courageous journalists around the world. I've watched the rise of the internet with a mixture of regard for emerging citizen journalism and worry over the disruptions of journalistic ethics, revenue streams, and community trust. All of these interests and concerns motivate this book.

This book began in another way with an invitation from Susan Moffitt and Brown University's Taubman Center for American Politics and Policy to give the Alexander Meiklejohn Lecture in 2018, and I am grateful to Susan and to the assembled group for invaluable comments and encouragement. Dean Madeleine Landrieu invited me to the Loyola Law School (New Orleans), where student law review editors polished and published the lecture; President Kathleen McCartney gave me the opportunity to engage with the Smith College community on these issues as part of a Constitution Day celebration. Yet the book began another way: in conversations during my entire life with my family, and most especially with my father, Newton Minow, whose life's work and searching commitment to the public interest inspire the questions and the hope for democracy and the quality of news media it and its members require. My mother and sisters also ensured that our household would be immersed in texts and art, devoted to public broadcasting, involved in media governance, and avidly engaged in critiquing what we saw and heard. When our parents sat us down to watch and then give reviews of early forms of educational television and films borrowed from Encyclopedia Britannica, they inspired their children to become a film critic, a teacher, and a leader of libraries (and lawyers!) with perspectives informing both this book and its author.

Harvard Law School colleagues have offered crucial support, challenging questions, and valuable advice. Special thanks—and no blame for my mistakes or missteps—go to Yochai Benkler, Susan Crawford, Einer Elhauge, Urs Gasser, Vicki Jackson, Randy Kennedy, Daphna Renan, Laura Weinrib, Jonathan Zittrain, and my dean and friend, John Manning. I am grateful to the Northwestern Pritzker School of Law for the opportunity to present some of this work, and notably to Doreen Gay Weisenhaus for suggestions. Jack Balkin, Sandra Baron, RonNell Anderson Jones, and Scott Shapiro offered vigorous comments, discussion, and suggestions under the auspices of Yale University's Information Society Project, and this book and my thinking are sharper as a result.

When I began the project, I served on the board of the CBS Corporation, which I left as it became the ViacomCBS Corporation, and also the board of the public broadcaster GBH, where I still serve. These experiences, plus my work in journalism at Northwestern University's Medill School, the once beloved *Chicago Daily News*, and the *Michigan Daily* help inform my work. E. J. Dionne, Archon Fung, Nancy Gibson, and Ann Marie Lipinski, each vital contributors to vibrant journalism and democratic accountability, involved me in discussions at once stimulating, concerning, and encouraging. Ann Marie Lipinski and the Nieman Fellows 2019–20 and alumni fellows—especially Carrie Johnson, Blair Kamin, Mary Ellen Klas, and Nathan Payne—gave me the gifts of their time and deep knowledge. So did Lowell Bergman, Tom Wheeler, Nicco Mele, Doug Smith, and Sabeel Rahman. Ken Goldstein and Elliot Schrage offered discerning comments as well as rich knowledge. I am grateful for their expertise and encouragement. Julius Genachowski, Nick Lehman, David McGraw, Deb Roy, David Rhodes, and Zia Rahman shared their perspectives and concerns about media, news, and democracy. Judy Smith generously went out of her way to offer comments, sources, and her work and the work of her students as I grappled with American culture and the history of media. Kristin Bumiller, Alice Hearst, Susan Silbey, Martha Umphrey, and Pat Williams offered comments and sustenance precisely when most needed. And thanks to Cynthia Dwork, Shafi Goldwasser, Jane Lipson, and Patricia Williams, I am learning about algorithmic tools and the challenges of harnessing them in servicc of humanity and human rights, while also finding and creating cheer during the pandemic.

I am grateful that Geoff Stone invited me to propose a book for this series; his encouragement of this project, wise advice, disagreements, and thorough editorial comments made the book and the process of writing it better in every way. And my gratitude to David McBride, Cheryl Merritt, and the Oxford University Press team is deep and enduring.

One of the greatest privileges of my life has been the chance to work with fabulous students. So many over the past four years have brought expertise, passion, and patience in my classes and with research, checking sources, and editing. My special gratitude goes to Akua Abu, Matthew Arons, Jeremy Dang, William Feldman, Kim Foreiter, Ben Horton, Jess Hui, Hannah Kannegieter, Hannah Klain, Jesse Lempel, Grace McLaughlin, Sarah Rutherford, Sejal Singh, Stephanie Sofer, Hannah Solomon-Strauss, Matthew Summers, and Yi Yuan, who each provided detailed research memoranda and comments on my work, and to Adira Levine, who did that and more. My undergraduate Radcliffe undergraduate partners Carissa Chen, Isabel Espinosa, Blessing Jee, Hilda Jones, Flora Li, and Luke Minton offered valuable insights and enthusiasm. First as a student and now as a colleague, evelyn douek has offered research, analyses, and more, saving me on many occasions from mistakes and pressing me rightly with her knowledge and powerful thinking about digital platforms and the global frontiers of freedom of speech; this is a much better work as a result. Our shared students generated fortifying ideas and enthusiasm. And I learned from teaching together with my former student Kendra Albert in ways that will be challenging and replenishing for years to come. Remaining mistakes are my own!

At the start, Rachel Keeler, and later Joseph Cunningham and Ellie Benagh, assisted administratively with this project and with the rest of my work; I am so thankful for their clarity, alacrity, and good cheer. Over and over, Larry Blum, Mary Casey, and Rick

Weissbourd have offered insights, moral support, and rescue from sad events. And how lucky I am to have Ken Greenberg as book-writing buddy; his insightful perceptions, historical acumen, and continuing encouragement were such a help, especially as the solitude of writing turned into the global solitude of social distancing. Also special thanks to Andrew Heyward, Jill Abramson, Eric Lander, Craig LeMay, Margie Marshall, Ellen Hume, and John Shattuck for conversations and for their examples as warriors for truth, journalism, and constitutional rights. Lucky for me, Judi Greenberg, Ken Greenberg, Gish Jen, Randy Kennedy, Jane Lipson, Chris Marks, Joe Marks, Elizabeth Yong, and many Singers and Adlands provide sustaining friendship.

Jo Minow, Nell Minow, Newton Minow, Mary Minow, and David Apatoff, thank you for believing in me. Newton Minow, thank you also for challenging me on every page, for sharing sources, for cheering on this project, for your pioneering and visionary efforts to realize the promise of modern communications technologies, and for your eloquent and far-seeing preface. You once told me that the laws governing communications touch every person directly, intimately, and daily, more so than perhaps any other laws; that and many other ways you teach me inspired this book. Mira Singer provided indispensable and meticulous editorial assistance as well as astonishing knowledge of the social media and pop culture worlds and indefatigable documentation of and protest against current national and global challenges. Joe Singer's suggestions were invariably persuasive; his kindness was constantly renewing. Mira and Joe, thanks for putting up with me!

We live in a time when human creations risk reshaping our worlds in ways that depart from the intention of individuals. This is, as poet Amanda Gorman says, the "sea we must wade." This is a time that calls for collective efforts to protect individual dignity, interpersonal respect, and reasonable prospects for seeking truth and for an imperfect but still possible constitutional democracy. Civil society, governments, researchers, and private enterprises can offer powerful checks on concentrated power, whether governmental and private. Digital natives born since the arrival of the internet see the promise and the perils of these times and offer navigational tools and dreams to guide older generations. May we collaborate and have the faith and energy to scaffold the climb ahead.

Notes

. . .

INTRODUCTION

1. U.S. Constitution, Amendment I. The U.S. Constitution ensures not only freedom of speech in general but also constitutional protection for the press. See Potter Stewart, "Or of the Press," *Hastings Law Journal* 26, no. 3 (1975): 631–637.

2. See Nicholas Lemann, "Can Journalism Be Saved?," *New York Review of Books*, Feb. 27, 2020, www.nybooks.com/articles/2020/02/27/can-journalism-be-saved.

3. "Losing the News: The Decimation of Local Journalism and the Search for Solutions," PEN America, Nov. 20, 2019, https://pen.org/wp-content/uploads/2019/12/Losing-the-News-The-Decimation-of-Local-Journalism-and-the-Search-for-Solutions-Report.pdf.

4. Penelope Muse Abernathy, "The Rise of a New Media Baron and the Emerging Threat of News Deserts," Center for Innovation and Sustainability in Local Media, University of North Carolina, 2016, http://newspaperownership.com/wp-content/uploads/2016/09/07.UNC_RiseOfNewMediaBaron_SinglePage_01Sep2016-REDUCED.pdf. More than sixty-five million Americans live in counties with only one local newspaper, or none at all. Clara Hendrickson, "Local Journalism in Crisis: Why America Must Revive Its Local Newsrooms," Brookings Institution, Nov.

12, 2019, https://www.brookings.edu/research/local-journalism-in-crisis-why-america-must-revive-its-local-newsrooms/.

5. Philip M. Napoli, Matthew Weber, Katie McCollough, and Qun Wang, "Assessing Local Journalism: News Deserts, Journalism Divides, and the Determinants of the Robustness of Local News," Duke Sanford School of Public Policy, Aug. 2018, https://dewitt.sanford.duke.edu/wp-content/uploads/2018/08/Assessing-Local-Journalism_100-Communities.pdf. The trend has continued. As Napoli and co-authors report, newspapers have lost 47 percent of their newsroom staff since 2004; at least two hundred counties (accounting for a total of over three million people) have no newspaper at all, and more than fifteen hundred counties have only one newspaper.

6. Megan Brenan, "Americans' Trust in Mass Media Edges Down to 41 Percent," Gallup, Sept. 26, 2019, https://news.gallup.com/poll/267047/americans-trust-mass-media-edges-down.aspx.

7. Knight Foundation, American Views 2020: Trust, Media, and Democracy (Aug. 4, 2020), https://knightfoundation.org/reports/american-views-2020-trust-media-and-democracy/

8. Julie Bosman, "How the Collapse of Local News Is Causing a 'National Crisis,'" *New York Times*, Nov. 20, 2019, https://www.nytimes.com/2019/11/20/us/local-news-disappear-pen-america.html; Hendrickson, "Local Journalism in Crisis."

9. Kristen Hare, "Here Are the Newsroom Layoffs, Furloughs and Closures Caused by the Coronavirus," Poynter, Apr. 6, 2020, https://www.poynter.org/business-work/2020/here-are-the-newsroom-layoffs-furloughs-and-closures-caused-by-the-coronavirus/.

10. Shelia Dang, "Google, Facebook Have Tight Grip on Growing U.S. Online Ad Market: Report," Reuters, June 5, 2019, https://www.reuters.com/article/us-alphabet-facebook-advertising/google-facebook-have-tight-grip-on-growing-u-s-online-ad-market-report-idUSKCN1T61IV. In 2018, Google and Facebook had an almost 60 percent market share of Internet ads, a 3 percent increase over the prior year.

11. Douglas McClennan and Jack Miles, "A Once Unimaginable Scenario: No More Newspapers," *Washington Post*, Mar. 21, 2018, https://www.washingtonpost.com/news/theworldpost/wp/2018/03/21/newspapers/.

12. See the Virginia Declaration of Rights, June 12, 1776, in *The Papers of George Mason Papers 1725–1792*, edited by Robert A. Rutland (Chapel

Hill: University of North Carolina Press, 1970), 1:287–289, press-pubs.uchicago.edu/founders/documents/v1ch1s3.html; Sam Lebovic, *Free Speech and Unfree News: The Paradox of Press Freedom in America* (Cambridge, MA: Harvard University Press, 2016), 9–13.

13. U.S. Constitution, Amendment I.

14. See generally Paul Starr, *The Creation of the Media: Political Origins of Modern Communications* (New York: Basic Books, 2004).

15. See Kyle Langvardt, "A New Deal for the On-Line Public Sphere," *George Mason Law Review* 26, no. 1 (2018): 53 ("Over the long run, it is hard to believe that political institutions in a well-functioning democracy would ignore problems such as the implosion of the publishing industry and its sudden and near-total replacement with a communications infrastructure that is primed for polarization, conspiracism, and sabotage. At a minimum, one would think that a well-functioning democracy might consider some form of product regulation to protect children from addictive products that are said to endanger their mental health and that obviously interfere with their education").

16. See, e.g., Sonya Fatah, "Charting the Evolution of Canadian News Consumption," Media in Canada, Dec. 15, 2015, https://mediaincanada.com/tag/ken-goldstein.

17. See Amartya Sen, *Development as Freedom* (Oxford: Oxford University Press, 1999), 16, 152–153.

18. Geoffrey R. Stone, *Perilous Times: Free Speech in Wartime* (New York: Norton, 2004).

19. Michael Luo, "The Fate of the News in the Age of the Coronavirus: Can a Fragile Media Ecosystem Survive the Pandemic?," *New Yorker*, Mar. 29, 2020, https://www.newyorker.com/news/annals-of-communications/the-fate-of-the-news-in-the-age-of-the-coronavirus.

CHAPTER ONE

1. See generally Nic Newman et al., *Reuters Institute Digital News Report 2017* (Oxford: Reuters Institute for the Study of Journalism, 2017), https://reutersinstitute.politics.ox.ac.uk/sites/default/files/Digital%20News%20Report%202017%20web_0.pdf. The 2019 version of the report

notes that 11 percent of those surveyed paid for their online news. Over half (55 percent) of those surveyed preferred to access news through an online algorithm rather than through human editors. Rasmus Kleis Nielsen et al., *Reuters Institute Digital News Report 2019* (Oxford: Reuters Institute for the Study of Journalism, 2019), https://reutersinstitute.politics.ox.ac.uk/sites/default/files/2019-06/DNR_2019_FINAL_0.pdf.

2. Catherine Buni, "4 Ways to Fund—and Save—Local Journalism," Nieman Reports, May 7, 2020, https://niemanreports.org/articles/4-ways-to-fund-and-save-journalism; Clara Hendrickson, "Critical in a Public Health Crisis, COVID-19 Has Hit Local Newsrooms Hard," Brookings Institution, Apr. 8, 2020, https://www.brookings.edu/blog/fixgov/2020/04/08/critical-in-a-public-health-crisis-covid-19-has-hit-local-newsrooms-hard.

3. Margaret Sullivan, *Ghosting the News: Local Journalism and the Crisis of American Democracy* (New York: Columbia Global Reports, 2020); See David Bodney, Local News is Struggling to Survive, AZCentral (Dec. 29, 2020), https://www.azcentral.com/story/opinion/op-ed/2020/12/29/local-news-essential-yet-struggling-survive/4064284001.

4. Penelope Muse Abernathy, "The Rise of a New Media Baron and the Emerging Threat of News Deserts," Center for Innovation and Sustainability in Local Media, University of North Carolina, 2016, http://newspaperownership.com/wp-content/uploads/2016/09/07.UNC_RiseOfNewMediaBaron_SinglePage_01Sept2016-REDUCED.pdf.

5. Elizabeth Grieco, "U.S. Newspapers Have Shed Half of Their Newsroom Employees Since 2008," Pew Research Center, Apr. 20, 2020, https://www.pewresearch.org/fact-tank/2020/04/20/u-s-newsroom-employment-has-dropped-by-a-quarter-since-2008; Hendrickson, "Critical in a Public Health Crisis." The Pew Research Center reported that between 2006 and 2018, circulation of daily newspapers in the United States and newsroom employment fell by nearly half. Pew Research Center, "Newspapers Fact Sheet," July 9, 2019, https://www.journalism.org/fact-sheet/newspapers.

6. Marc Tracy, "News Media Outlets Have Been Ravaged by the Pandemic," *New York Times*, Apr. 10, 2020, https://www.nytimes.com/2020/04/10/business/media/news-media-coronavirus-jobs.html. The paper updated its initial report of 33,000 news media jobs affected to 36,000 by May 1, 2020.

7. Clara Hendrickson, "Local Journalism in Crisis: Why America Must Revive Its Local Newsrooms," Brookings Institution, Nov. 12, 2019, https://www.brookings.edu/research/local-journalism-in-crisis-why-america-must-revive-its-local-newsrooms.

8. State of Public Trust in Local News, Knight Report/Gallup, 2019, 2, https://kf-site-production.s3.amazonaws.com/media_elements/files/000/000/440/original/State_of_Public_Trust_in_Local_Media_final_.pdf: "Local news bests national news in earning more trust of Americans for coverage that Americans can use in their daily life (79 percent to 19 percent) and in reporting without bias (66 percent to 31 percent), among other roles and responsibilities." See also John Sands, "Local News Is More Trusted Than National News but That Could Change," Knight Foundation, Oct. 29, 2019, https://knightfoundation.org/articles/local-news-is-more-trusted-than-national-news-but-that-could-change.

9. Paul Moses, "In New York City, Local Coverage Declines—and Takes Accountability with It," *Daily Beast*, Apr. 3, 2017, https://www.thedailybeast.com/in-new-york-city-local-coverage-declinesand-takesaccountability-with-it.

10. Richard Fausset, "Dying Gasp of One Local Newspaper," *New York Times*, Aug. 1, 2019, https://www.nytimes.com/2019/08/01/uw/warroad-pioneer-news-desert.html?module-inline; John Enger, "After 121 Years, the Warroad Pioneer Newspaper Closes Its Doors," NPR, May 8, 2019, https://www.mprnews.org/story/2019/05/08/after-121-years-the-warroad-pioneer-closes-its-doors.

11. See Stephen Deere, Chuck Raasch, and Jeremy Kohler, "DOJ Finds Ferguson Targeted African-Americans, Used Courts Mainly to Increase Revenue," *St. Louis Post-Dispatch*, Mar. 5, 2015, https://www.stltoday.com/news/local/crime-and-courts/doj-findsferguson-targeted-african-americans-used-courts-mainly-to/article_d561d303-1fe556b7-b4ca-3a5cc9a75c82.html. On the centrality of political accountability to the First Amendment, see Gia Lee, "Persuasion, Transparency, and Government Speech," *Hastings Law Journal* 56, no. 5 (2005): 983–1057.

12. Josh Wilson, "Fixing Journalism's Ability to Promote Civic Good Should Be the Focus of Philanthropic Giving," *Chronicle of Philanthropy*, Nov. 1, 2017, https://www.philanthropy.com/article/Opinion-Fixing-Journalism-s/241610.

13. See generally Douglas Brinkley, *Cronkite* (New York: Harper, 2012), 2, 5, 663–666.

14. Brent Cunningham, "Re-Thinking Objectivity," *Columbia Journalism Review*, July/Aug. 2003, https://archives.cjr.org/feature/re-thinking_objectivity.php. For a more skeptical view, see C. W. Anderson, Leonard Downie Jr., and Michael Schudson, *The News Media: What Everyone Needs to Know* (New York: Oxford University Press, 2016), 55–57.

15. Robert Kaiser, "The Bad News About the News," Brookings Institution, Oct. 16, 2014, http://csweb.brookings.edu/content/research/essays/2014/badnews.html.

16. Jim Waterson, "Influencers Among 'Key Distributors' of Coronavirus Misinformation," *The Guardian*, Apr. 8, 2020, https://www.theguardian.com/media/2020/apr/08/influencers-being-key-distributors-of-coronavirus-fake-news.

17. See Damian Radcliffe and Christopher Ali, "Local News in a Digital World: Small Market Newspapers in the Digital Age," Columbia University, 2017, https://academiccommons.columbia.edu/doi/10.7916/D8WS95VQ.

18. Communications Management Inc., "Research Note Supporting Canadian Journalism," Apr. 24, 2019, http://media-cmi.com/downloads/CMI_Research_Note_Supporting_Canadian_Journalism_042419.pdf.

19. Brad Adgate, "Newspaper Revenue Drops as Local News Interest Rises Amid Coronavirus," *Forbes*, Apr. 13, 2020, https://www.forbes.com/sites/bradadgate/2020/04/13/newspapers-are-struggling-with-coronavirus/#2221b5a339ef; see also Elizabeth Grieco, "U.S. Newspapers Have Shed Half of Their Newsroom Employees Since 2008," Pew Research Center, Apr. 20, 2020, https://www.pewresearch.org/fact-tank/2020/04/20/u-s-newsroom-employment-has-dropped-by-a-quarter-since-2008.

20. "How Leading American Newspapers Got People to Pay for News," *The Economist*, Oct. 26, 2017, https://www.economist.com/news/business/21730683first-three-part-series-future-journalism-how-leading-american-newspapers-go.

21. Newman et al., *Reuters Institute Digital News Report 2017*.

22. Roy Greenslade, "Almost 60 Percent of US Newspaper Jobs Vanish in 26 Years," *The Guardian*, June 6, 2016, https://www.theguardian.com/media/greenslade/2016/jun/06/almost-60-of-us-newspaper-jobs-vanish-in-26-years.

23. "Buying Spree Brings More Local TV to Fewer Big Companies," Pew Research Center, May 11, 2017, https://pewresearch.org/fact-tank/2017/05/11/buying-spree-brings-more-local-tv-stations-to-fewer-big-companies.

24. Ann Marie Lipinski, "Journalism Needs a Strategy to Bridge the Chasm Between the Now and the Next," Nieman Reports, Feb. 2020, https://niemanreports.org/articles/journalism-needs-a-strategy-to-bridge-the-chasm-between-the-now-and-the-next (quoting John Williams, talk show host, on threatened takeover of the *Chicago Tribune* by Alden Global Capital, a New York hedge fund "with a reputation for more newsroom strip-mining than investing").

25. See Robert W. McChesney and John Nichols, *The Death and Life of American Journalism: The Media Revolution That Will Begin the World Again* (Philadelphia: Nation Books, 2010), 38–41, 47–49, 61–63.

26. See Jack Shafer, "Why Newspapers Have Gone to Hell," *Slate*, June 27, 2011, http://www.slate.com/articles/news_and_politics/press_box/2011/06/why_newspapers_ have_gone_to_hell.html; see generally Katharine Q. Seelye and Andrew Ross Sorkin, "Knight Ridder Newspaper Chain Agrees to Sale," *New York Times*, Mar. 12, 2006, http://www.nytimes.com/2006/03/12/archives/knight-ridder-newspaper-chain-agreesto-sale.html.

27. The pattern of declining revenues, bankruptcy, and sale to a hedge fund is illustrated by what happened to a family-run newspaper chain. See Marc Tracy, "McClatchy, Family-Run News Chain, Goes to Hedge Fund in Bankruptcy Sale," *New York Times*, Aug. 4, 2020, https://www.nytimes.com/2020/08/04/business/media/mcclatchy-newspapers-bankrutpcy-chatham.html.

28. See generally Eli Noam, "Media Concentration in the United States," in *Who Owns the World's Media? Media Concentration and Ownership Around the World* (New York: Oxford University Press, 2016); *See* Damian Radcliffe and Christopher Ali, *Local News in a Digital World: Small Market Newspapers in the Digital Age.* Tow Center for Digital Journalism (A Tow/Knight Report Fall 2017), available at COLUM. ACAD. COMMONS (2017).

29. "Losing the News: The Decimation of Local Journalism and the Search for Solutions," PEN America, Nov. 20, 2019, https://pen.org/wp-content/uploads/2019/12/Losing-the-News-The-Decimation-of-Local-Journalism-and-the-Search-for-Solutions-Report.pdf.

30. See generally Gene Roberts, ed., *Leaving Readers Behind: The Age of Corporate Newspapering* (Fayetteville: University of Arkansas Press, 2001).

31. See Nic Newman et al., *Reuters Institute Digital News Report 2018* (Oxford: Reuters Institute for the Study of Journalism, 2018), 14–15, https://agency.reuters.com/content/dam/openweb/documents/pdf/news-agency/report/dnr-18.pdf.

32. See Carina Tenor, "Hyperlocal News and Media Accountability," *Digital Journalism* 6, no. 8 (2018): 1064–1077, DOI: 10.1080/21670811.2018.1503059; Sara Fischer and Scott Rosenberg, "Power Pendulum Swings Back to News Companies," Axios, Apr. 2, 2019, https://www.axios.com/news-power-pendulum-swings-back-1554162971-6bc6e37f-db87-4ed0-9d87-3091ac868c33.html.

33. RonNell Andersen Jones and Sonja R. West, "The Fragility of the Free American Press," *Northwestern University Law Review* 112 (2017): 48, 56 (quoting Jordan Weissmann, "The Decline of Newspapers Hits a Stunning Milestone," *Slate*, Apr. 28, 2014, https://slate.com/business/2014/04/decline-of-newspapers-hits-a-milestone-print-revenue-is-lowest-since-1950.html).

34. See generally Sasha Chavkin, "The Koch Brothers Media Investment (Updated)," *Columbia Journalism Review*, Apr. 22, 2013, https://archives.cjr.org/united_states_project/the_koch_brothers_media_invest.php; Matt Gertz, "Breitbart Is Not Independent: It's the Communications Arm of the Mercers' Empire," *Salon*, Apr. 24, 2017, https://www.salon.com/2017/04/24/breitbartis-not-independent-its-the-communications-arm-of-the-mercers-empire_partner; David McKnight, "Rupert Murdoch's News Corporation: A Media Institution with a Mission," *Historical Journal of Film, Radio and Television* 30 (2010): 303–316; Jim Bucknell, "Ideology Runs Rampant at Rupert Murdoch's Australian Newspaper," *The Guardian*, Dec. 7, 2015, https://www.theguardian.com/commentisfree/2015/dec/07/ideology-runs-rampant-at-rupert-murdochs-australian-newspaper.

35. Andrew Ross Sorkin, "Peter Thiel, Tech Billionaire, Reveals Secret War with Gawker," *New York Times*, May 25, 2016, https://www.nytimes.com/2016/05/26/business/dealbook/peter-thiel-tech-billionaire-reveals-secret-war-with-gawker.html; Sydney Ember, "Gawker, Filing for Bankruptcy After Hulk Hogan Suit, Is for Sale," *New York Times*, June 10, 2016, https://www.nytimes.com/2016/06/11/business/media/gawker-bankruptcy-sale.html.

36. Eli Rosenberg, "Trump Said Sinclair 'Is Far Superior to CNN.' What We Know About the Conservative Media Giant," *Washington Post*, Apr. 3, 2018, https://www.washingtonpost.com/news/style/wp/2018/04/02/

get-to-know-sinclairbroadcast-group-the-conservative-local-news-giant-with-a-growingreach/?noredirect=on&utm_term=.6a58d1478d22.

37. Lucia Graves, "This Is Sinclair: 'The Most Dangerous US Company You've Never Heard Of,'" *The Guardian*, Aug. 17, 2017, https://www.theguardian.com/media/2017/aug/17/sinclair-news-media-fox-trumpwhite-house-circa-breitbart-news; Robert Channick, "Under Sinclair, WGN Would Be Chicago's 'Very Own' No More," *Chicago Tribune*, Aug. 10, 2017, http://www.chicagotribune.com/business/ct-tribune-sinclair-merger-wgn-0813-biz20170810-story.html; John D. McKinnon, "Sinclair Broadcasting Agrees to Pay Record Penalty to End FCC Probes," *Wall Street Journal*, May 6, 2020, https://www.wsj.com/articles/sinclair-broadcasting-agrees-to-pay-record-penalty-to-end-fcc-probes-11588809191.

38. Erika Fry, "Briefing: Super Bowl Ads Can't Save TV," *Fortune*, Feb. 1, 2018, 11, 12 (relying on an analysis of Nielsen data by MarketingCharts.com).

39. Erik Barnouw, *The Golden Web: A History of Broadcasting in the United States, 1933–1953* (New York: Oxford University Press, 1968), 2:77–78; Charles L. Ponce de Leon, *That's the Way It Is: A History of Television News in America* (Chicago: University of Chicago Press, 2015), ix–xvii, 4–5; Jack Mirkinson, "60 Years Ago, Edward R. Murrow Took Down Joseph McCarthy," Huffington Post, Mar. 10, 2014, https://www.huffingtonpost.com/2014/03/10/edward-murrow-joseph-mccarthy-60-years-later_n_4936308.html.

40. Frank Pasquale, *The Black Box Society: The Secret Algorithms That Control Money and Information* (Cambridge, MA: Harvard University Press, 2015), 8.

41. Sarah Perez, "Plex Adds Personalized, Streaming News to Its Media Player Software," TechCrunch, Sept. 26, 2017, https://techcrunch.com/2017/09/26/plex-adds-personalized-streaming-news-to-its-media-player-software; see also Casey Newton, "Google Introduces the Feed, a Personalized Stream of News on iOS and Android," The Verge, July 19, 2017, https://www.theverge.com/2017/7/19/15994156/google-feed-personalized-news-stream-android-ios-app; Yukinori Koide, "Optimize Delivery of Trending, Personalized News Using Amazon Kinesis and Related Services," Amazon Web Services, Jan. 18, 2018, https://aws.amazon.com/blogs/bigdata/optimize-delivery-of-trending-personalized-news-using-amazon-kinesis-andrelated-services.

42. Dan Price, "5 Free Streaming News Channels for Cord-Cutters," Make Use Of, May 18, 2017, https://www.makeuseof.com/tag/free-streaming-news-channels-cord-cutters.

43. "Facebook Reports Fourth Quarter and Full Year 2019 Results," PR Newswire, Jan. 29, 2020, https://www.prnewswire.com/news-releases/facebook-reports-fourth-quarter-and-full-year-2019-results-300995616.html.

44. Timothy Stenovec, "Facebook Is Now Bigger Than the Largest Country on Earth," Huffington Post, Jan. 28, 2015, https://www.huffingtonpost.com/2015/01/28/facebook-biggest-country_n_6565428.html.

45. Emily Bell, "Why Facebook's News Feed Changes Are Bad News for Democracy," *The Guardian*, Jan. 21, 2018, https://www.theguardian.com/media/mediablog/2018/jan/21/why-facebook-news-feed-changes-bad-news-democracy.

46. See Lili Levi, "A 'Faustian Pact'? Native Advertising and the Future of the Press," *Arizona Law Review* 57 (2015): 647.

47. Tom Wheeler, "How to Monitor Fake News," *New York Times*, Feb. 20, 2018, https://www.nytimes.com/2018/02/20/opinion/monitor-fake-news.html?emc=edit_th_180221&nl=todaysheadlines&nlid=378183740221editorical.

48. Frank Pasquale, "The Automated Public Sphere," University of Maryland Francis King Carey School of Law, Legal Studies Research Paper No. 2017-31 (2017), 4–6, https://papers.ssrn.com/sol3/papers.cfm?abstract_id=3067552. A group of psychologists have called the use of psychological techniques to hook users on sites unethical as it goes beyond engagement into compulsion and addiction. Chavie Lieber, "Tech Companies Use 'Persuasive Design' to Get Us Hooked. Psychologists Say it's Unethical," Vox (Aug. 8, 2019), https://www.vox.com/2018/8/8/17664580/persuasive-technology-psychology.

49. Shona Ghosh, "Sheryl Sandberg Just Dodged a Question About Whether Facebook Is a Media Company," *Business Insider*, Oct. 12, 2017, http://www.businessinsider.com/sheryl-sandberg-dodged-question-on-whetherfacebook-is-a-media-commpany-2017-10; Kirsten Grind, Sam Schechner, Robert McMillan, and John West, "How Google Interferes with Its Search Algorithms and Changes Your Results," *Wall Street Journal*, Nov. 15, 2019, https:www.wsj,com/articles/how-google-interferes-with-its-search-algorithms-and-changes-your-results-11573823753.

50. Katharine Viner, "How Technology Disrupted the Truth," *The Guardian*, July 12, 2016, https://www.theguardian.com/media/2016/jul/12/how-technologydisrupted-the-truth (quoting Emily Bell, director of the Tow Center for Digital Journalism at Columbia University's Graduate School of Journalism).

51. Nicole Nguyen, "Doomscrolling: Why We Can't Just Look Away," *Wall Street Journal*, June 7, 2020, https://www.wsj.com/articles/doomscrolling-why-we-just-cant-look-away-11591522200.

52. Kevin Roose, Mike Isaac, and Sheera Frenkel, "The Clash at Facebook: Pragmatists vs. ldealists," *New York Times*, Nov. 25, 2020.

53. See Evan Osnos, "Ghost in the Machine: Can Mark Zuckerberg Fix Facebook Before It Breaks Democracy?," *New Yorker*, Sept. 17, 2018, 32, 41–44, 46–47.

54. Lauren Etter, "What Happens When the Government Uses Facebook as a Weapon?," Bloomberg, Dec. 7, 2017, https://www.bloomberg.com/news/features/2017-12-07/how-rodrigo-duterte-turnedfacebook-into-a-weapon-with-a-little-help-from-facebook. The traditional media routinely identifies Rodrigo Duterte as an autocrat with oppressive practices. See, e.g., "Philippine Strongman Duterte Has Some Thoughts on Condoms," Marketwatch, Feb. 20, 2018, https://www.marketwatch.com/story/philippine-strongman-duterte-has-some-view-on-condoms-2018-02-17.

55. See Zeynep Tufekci, "Zuckerberg's Preposterous Defense of Facebook," *New York Times*, Sept. 29, 2017, https://www.nytimes.com/2017/09/29/opinion/mark-zuckerbergfacebook.html.

56. See *Gonzalez v. Google Inc.*, 282 F. Supp. 3d 1150 (N.D. Ca. 2017); Alexis Kramer, "Twitter, Facebook, Google Not Liable for Hamas Posts," Bloomberg, Dec. 6, 2017, https://www.bna.com/twitter-facebook-google-n73014472792.

57. Tarlton Gillespie, *Custodians of the Internet: Platforms, Content Moderation, and the Hidden Decisions That Shape Social Media* (New Haven: Yale University Press, 2018), 13.

58. Mike Isaac and Sheera Frenkel, "Facebook Is 'Just Trying to Keep the Lights On' as Traffic Soars in Pandemic," *New York Times*, Mar. 24, 2020, https://www.nytimes.com/2020/03/24/technology/virus-facebook-usage-traffic.html. In the midst of COVID-19, Facebook shifted from leaving fact-checking to users to including warnings about misinformation in the news feeds of users exposed to misinformation and redirection to a

page created by the World Health Organization to counter myths about the virus. Casey Newton, "Coronavirus Misinformation Is Putting Facebook to the Test," The Verge, Apr. 17, 2020, https://www.theverge.com/interface/2020/4/17/21223742/coronavirus-misinformation-facebook-who-news-feed-message-avaaz-report.

59. Yoel Roth and Nick Pickles, "Updating Our Approach to Misleading Information," Twitter, May 11, 2020, https://blog.twitter.com/en_us/topics/product/2020/updating-our-approach-to-misleading-information.html.

60. See Jeff Horwitz, "Facebook Is Doing Too Little on Civil-Rights Concerns, Auditors Say," *Wall Street Journal*, July 9, 2020, https://www.wsj.com/articles/facebook-is-doing-too-little-on-civil-rights-concerns-auditors-say-11594198537.

61. Camila Domonoske, "Students Have 'Dismaying' Inability to Tell Fake News from Real, Study Finds," NPR, Nov. 23, 2016, https://www.npr.org/sections/thetwo-way/2016/11/23/503129818/study-finds-students-have-dismaying-inability-to-tell-fake-news-from-real.

62. Melissa de Zwart, "Keeping the Neighbourhood Safe: How Does Social Media Moderation Control What We See (and Think)?," *Alternative Law Journal* 43 (2018): 283.

63. Scott Shane and Vindu Goel, "Fake Russian Facebook Accounts Bought $100,000 in Political Ads," *New York Times*, Sept. 6, 2017, https://www.nytimes.com/2017/09/06/technology/facebook-russian-political-ads.html; Mark Mazzetti and Katie Benner, "12 Russian Agents Indicted in Mueller Investigation," *New York Times*, July 13, 2018, https://www.nytimes.com/2018/07/13/us/politics/mueller-indictment-russian-intelligence-hacking.html; John Marshall, "Facebook Still Lying About Its Role in the 2016 Election," Talking Points Memo, Feb. 17, 2018, https://talkingpointsmemo.com/edblog/facebook-still-lying-aboutits-role-in-the-2016-election.

64. Georgia Wells and Robert McMillan, "Facebook Battles New Criticism After U.S. Indictment Against Russians," *Wall Street Journal*, Feb. 19, 2018, https://www.wsj.com/articles/facebook-battles-new-criticism-after-u-s-indictmentagainst-russians-1519066080 (quoting Sam Wooley, Oxford research associate studying social media platforms). Russia apparently uses both disinformation and cyberattacks. See Megan Reiss, "Takeaways from the Latest Russian Hacking Indictment," Lawfare, Oct. 4, 2018, https://www.lawfareblog.com/takeaways-latestrussian-hacking-indictment. For further developments in federal investigations, see Adam

Goldman, "Justice Dept. Accuses Russians of Interfering in Midterm Elections," *New York Times*, Oct. 19, 2018, https://www.nytimes.com/2018/10/19/us/politics/russiainterference-midterm-elections.html.

65. Section 230 of the Communications Decency Act of 1996 (Title V of the Telecommunications Act of 1996), codified at 47 U.S.C. § 230, shields internet service providers and other intermediaries from liability for postings by third parties. Communications Decency Act of 1996, 47 U.S.C. § 230 (2012); see generally *Fair Housing Council v. Roommates.com, LLC*, 521 F.3d 1157 (9th Cir. 2008). Social media can also amplify gang violence by offering channels for targeting retribution or bragging about violence. See Desmond Upton Patton et al., "What's a Threat on Social Media? How Black and Latino Chicago Young Men Define and Navigate Threats Online," *Youth and Society* 51 (2017): 756, https://journals.sagepub.com/doi/10.1177/0044118X17720325.

66. Jeff Kosseff, *The Twenty-Six Words That Created the Internet* (Ithaca, NY: Cornell University Press, 2019), 1.

67. Compare Department of Justice, Office of Public Affairs, "The Department of Justice Unveils Proposed Section 230 Legislation," Sept. 23, 2020, https://www.justice.gov/opa/pr/justice-department-unveils-proposed-section-230-legislation, with "Nancy Pelosi Warns Tech Companies That Section 230 'Is in Jeopardy,'" TechCrunch, Apr. 12, 2019, https://techcrunch.com/2019/04/12/nancy-pelosi-section-230.

68. See *Fair Housing Council v. Roommates.com, LLC*, 521 F.3d 1157, 1169, 1173–74 (9th Cir. 2008); see also *Doe v. Internet Brands, Inc.*, 824 F.3d 846, 852–53 (9th Cir. 2016).

69. Joan Donavan, "Covid Hoaxes Are Using a Loophole to Stay Alive—Even After Content Is Deleted," *MIT Tech Review*, Apr. 20, 2020, https://www.technologyreview.com/2020/04/30/1000881/covid-hoaxes-zombie-content-wayback-machine-disinformation; Grind et al., "How Google Interferes with Its Search Algorithms"; John Jantsch, "Why Dark Posts Are the Best Facebook Advertising Approach Right Now," Duct Tape Marketing, https://www.ducttapemarketing.com/facebook-dark-posts, last visited Oct. 2, 2018. After the 2016 election, when the practice became more known, Facebook ended the "dark post" option. Garett Sloane, "No More 'Dark Posts': Facebook to Reveal All Ads," Ad Age, Oct. 27, 2017, https://adage.com/article/digital/facebook-drag-darkposts-light-election/311066.

70. Gillian Bolsover and Philip Howard, "Computational Propaganda and Political Big Data: Moving Toward a More Critical Research Agenda," *Big Data* 5 (Dec. 1, 2017): 273–276, https://doi.org/10.1089/big.2017.29024.cpr. Given evidence of organized social media manipulation campaigns in seventy countries in 2019, twenty-two more than in 2018, Facebook is the platform of choice for computational propaganda, with evidence of manipulation in fifty-six different countries. Samantha Bradshaw and Philip N. Howard, "The Global Disinformation Order: 2019 Inventory of Organised Social Media Manipulation," University of Oxford, 2019, https://comprop.oii.ox.ac.uk/wp-content/uploads/sites/93/2019/09/CyberTroop-Report19.pdf.

71. Alexis C. Madrigal, "What Facebook Did to American Democracy," *The Atlantic*, Oct. 12, 2017, https://www.theatlantic.com/technology/archive/2017/10/what-facebook-did/542502.

72. Eli Pariser, *The Filter Bubble: What the Internet Is Hiding from You* (London: Penguin, 2011); Cass Sunstein, *#Republic: Divided Democracy in the Age of Social Media* (Princeton, NJ: Princeton University Press, 2017).

73. Jonathan Zittrain, "Engineering an Election," *Harvard Law Review Forum* 127 (2014); Jonathan Zittrain, "Facebook Could Decide an Election Without Anyone Ever Finding Out," *New Republic*, June 1, 2014, https://newrepublic.com/article/117878/information-fiduciary-solution-facebook-digital-gerrymandering.

74. See, e.g., Richard Nieva, "Facebook Now Tells You Exactly Why It Takes Down Posts," CNET, Apr. 23, 2018, https://www.cnet.com/news/facebook-releasesinternal-guidelines-for-taking-down-posts; Ben Wolfgang, "Thumbs Down: Facebook's Hate Speech and Censorship Policies No Easy Fix," *Washington Times*, Apr. 22, 2018, https://www.washingtontimes.com/news/2018/apr/22/facebook-hate-speechcensorship-policies-upset-bot. Some critics note that Facebook does not alter its central focus on targeting individuals based on using their data trail. See Kari Paul, "Facebook to Remove 'Trending' News from Its Site Amid Fake News Criticism," Marketwatch, June 3, 2018, https://www.marketwatch.com/story/facebook-to-remove-trending-news-from-its-site-amid-fake-news-criticism-2018-06-01. Dystopic images of spreading hatred on social media are not works of fiction, but experiences unfolding in 2020. See, e.g., Mohammad Ali, "The Influencers," *Wired*, May 2020 (documenting hatred and terrorism campaigns on social media against Muslims in India).

75. Joel Flaxman, Sharad Goel, and Justin M. Rao, "Filter Bubbles, Echo Chambers, and Online News Consumption," *Public Opinion Quarterly* 80 (2016): 298.

76. Yochai Benkler, Robert Faris, and Hal Roberts, *Network Propaganda Manipulation, Disinformation, and Radicalization* (New York: Oxford University Press, 2018).

77. David Robson, "The Myth of the Online Echo Chamber," BBC, Apr. 16, 2018, https://www.bbc.com/future/article/20180416-the-myth-of-the-online-echo-chamber.

78. Todd Spangler, "Facebook Slams Netflix's 'The Social Dilemma' as 'Distorted' and Sensationalist," *Variety*, Oct. 2, 2020, https://variety.com/2020/digital/news/facebook-netflix-social-dilemma-documentary-1234791015.

79. Steven Waldman, "The Information Needs of Communities: The Changing Media Landscape in a Broadband Age," Federal Communications Commission, 2011, https://transition.fcc/gove/osp/inc-report/The_Information_Needs_of_Communities.pdf.

80. Pengjie Gao, Chang Lee, and Dermot, "Financing Dies in Darkness? The Impact of Newspaper Closures on Public Finance," *Journal of Financial Economics* 135, no. 2 (2020): 445, https://ssrn.com/abstract=3175555 http://dx.doi.org/10.2139/ssrn.3175555. In contrast, the availability of local accountability journalism changes the behavior of leaders. Charles Ornstein, "Local Accountability Journalism Still Has a Huge Impact," ProPublica, Jan. 30, 2020, https://www.propublica.org/article/local-accountability-journalism-still-has-a-huge-impact.

81. Joshua P. Darr, Matthew P. Hitt, and Johanna L. Dunaway, "Newspaper Closures Polarize Voting Behavior," *Journal of Communication* 68, no. 6 (Dec. 2018): 1007, https://academic.oup.com/joc/issue/68/6; Danny Hays and Jennifer L. Lawless, "The Decline of Local News and Its Effects: New Evidence from Longitudinal Data," *Journal of Politics* 80, no. 1 (Jan. 2018): 332, https://www.journals.uchicago.edu/doi/full/10.1086/694105?mobileUi=0&; Sarah Holder, "When Local Newsrooms Shrink, Fewer Candidates Run for Mayor," CityLab, Apr. 11, 2019, https://www.citylab.com/life/2019/04/local-news-decline-journalist-news-desert-california-data/586759; Lee Shaker, "Dead Newspapers and Citizens' Civic Engagement," *Political Communication* 31 (2014): 10, https://www.tandfonline.com/doi/full/10.1080/10584609.2012.762817.

82. Mathew Ingram, "Google and Facebook Account for Nearly All Growth in Digital Ads," Fortune, Apr. 26, 2017, http://fortune.com/2017/04/26/google-facebook-digitalads; see also Newman et al., *Reuters Institute Digital News Report* 2017, 101–103; Henri Gendreau, "Don't Stop the Presses! When Local News Struggles, Democracy Withers," *Wired*, Nov. 30, 2017, https://www.wired.com/story/dont-stop-the-presses-why-big-tech-should-subsidizereal-journalism.

83. Matthew Gentzkow, "Trading Dollars for Dollars: The Prices of Attention Online and Offline," *American Economic Review* 104 (May 2014): 481. Furthermore, spending on digital advertising has overtaken spending on ads for television and other media over the past ten years; Suzanne Vranica, "Facebook and Google Confront Antagonism of Big Advertisers," *Wall Street Journal*, Mar. 26, 2018, https://www.wsj.com/articles/facebook-and-google-face-emboldened-antagonists-bigadvertisers-1521998394. See Michael Barthel, "5 Key Takeaways about the State of the News Media in 2018," Pew Research Center, Jul. 23, 2019, https://www.pewresearch.org/fact-tank/2019/07/23/key-takeaways-state-of-the-news-media-2018.

84. Hendrickson, "Local Journalism in Crisis."

85. Rasmus Kleis Nielsen and Meera Selva, "More Important, but Less Robust? Five Things Everybody Needs to Know About the Future of Journalism," Reuters Institute, Jan. 2019, https://reutersinstitute.politics.ox.ac.uk/sites/default/files/2019-01/Nielsen_and_Selva_FINAL_0.pdf.

86. James Hohmann, "The Corrosion of Support for First Amendment Principles Started Before Trump. He's Supercharged It," *Washington Post*, Oct. 23, 2017, https://www.washingtonpost.com/news/powerpost/paloma/daily-202/2017/10/23/daily-202-the-corrosion-of-support-for-first-amendment-principles-startedbefore-trump-he-s-supercharged-it/59ed49b130fb045cba000926/?utm_term=.adf193d03d56.

87. Michelle Ye Hee Lee, "FEC Struggles to Craft New Rules for Political Ads in the Digital Space," *Washington Post*, June 28, 2018, https://www.washingtonpost.com/politics/fec-struggles-to-craft-new-rules-for-political-ads-in-the-digital-space/2018/06/28/c749a234-7af9-11e8-aeee-4d04c8ac6158_story.html?utm_term=.53fffe4b08c7; Ann Ravel, "How the FEC Turned a Blind Eye to Foreign Meddling," Politico, Sept. 18, 2017, https://www.politico.com/magazine/story/2017/09/18/fec-foreign-meddling-russiafacebook-215619; Kenneth P. Doyle, "Facebook Political Ads Get Bare-Bones Guidance from FEC," Bloomberg, Dec. 15, 2017, http://news.

bna.com/mpdm/MPDMWB/split_display.adp?fedfid=125137469&vname=mpebulallissues&jd=000001 605713dc0aa3657f3feaba0002&split=0.

88. Douglas MacMillan, "YouTube Says It Mistakenly Promoted a Conspiratorial Video on Florida Shooting," *Wall Street Journal*, Feb. 22, 2018, https://www.wsj.com/articles/youtube-says-it-mistakenly-promoted-a-conspiratorialvideo-on-florida-shooting-1519257359.

89. E.g., Brian Fung, "FCC Plan Would Give Internet Providers Power to Choose the Sites Customers See and Use," *Washington Post*, Nov. 21, 2017, https://www.washingtonpost.com/news/the-switch/wp/2017/11/21/the-fcc-has-unveiled-its-plan-to-rollback-its-net-neutrality-rules.

90. "Press Widely Criticized, but Trusted More Than Other Information Sources: Views of the Media: 1985–2011," Pew Research Center, 2011, http://www.people-press.org/2011/09/22/press-widely-criticized-but-trusted-morethan-other-institutions; Uri Friedman, "Trust Is Collapsing in America," *The Atlantic*, Jan. 21, 2018, https://www.theatlantic.com/international/archive/2018/01/trusttrump-america-world/550964.

91. James Carson, "Fake News: What Exactly Is It—And How Can You Spot It?," *The Telegraph*, Feb. 16, 2018, http://www.telegraph.co.uk/technology/0/fake-news-exactly-has-really-had-influence.

92. See generally Jennifer Kavanagh and Michael D. Rich, *Truth Decay: An Initial Exploration of the Diminishing Role of Facts and Analysis in American Public Life* (Santa Monica, CA: RAND, 2018).

93. See Paul Farhi and Sarah Ellison, "Ezra Klein Leaves Vox, the Website He Founded, for New York Times, in a Digital Media A-List Exodus," *Washington Post*, Nov. 20, 2020, https://www.washingtonpost.com/lifestyle/media/ezra-klein-vox-departure-digital-media/2020/11/20/289730ea-2b5d-11eb-92b7-6ef17b3fe3b4_story.html.

94. Cullen Murphy, "Predictions About the Internet Are Probably Wrong," *The Atlantic*, Jan.–Feb. 2020, https://www.theatlantic.com/magazine/archive/2020/01/before-zuckerberg-gutenberg/603034.

95. See Paul Bradshaw, "How the Web Changed the Economics of News—In All Media," Online Journalism Blog, June 4, 2009, https://onlinejournalismblog.com/2009/06/04/how-the-web-changed-the-economics-of-news-in-all-media (last updated Oct. 9, 2012).

96. See "Distributed Network," Techopedia, https://www.techopedia.com/definition/27788/distributed-network (last visited Dec. 31, 2018). Some describe the internet as a network of networks: Paul Tulenko, "Internet

Is a Network of Networks," *Deseret News*, May 9, 1995, https://www.deseretnews.com/article/419986/Internet-is-a-network-of-networks.html.

97. Jeff Tyson, "How Internet Infrastructure Works," How Stuff Works, Apr. 3, 2001, https://computer.howstuffworks.com/Internet/basics/Internet-infrastructure.htm; "How Do Mobile Phones and the Internet Work?," Me and My Shadow, last updated Mar. 4, 2016, https://myshadow.org/how-do-mobile-phones-and-Internet-work.

98. See Nils B. Weidmann et al., "Digital Discrimination: Political Bias in Internet Service Provision Across Ethnic Groups," *Science* 353 (Sept. 9, 2016): 1151, 1152, 1154, http://science.sciencemag.org/content/353/6304/1151.full?ijkey= 7Wq4RKNGjbIvw&keytype=ref&siteid=sci.

99. Dong Ngo, "Home Networking Explained, Part 4: Wi-Fi vs. Internet," CNET, Sept. 3, 2016, https://www.cnet.com/how-to/ homenetworking-explained-part-4-wi-fi-vs-Internet; Allie Shaw, "How Does Fiber Internet Work?," Reviews.org, Oct. 3, 2016, https://www.reviews.org/Internet-service/fiberInternet-work.

100. See Jonathan Taplin, *Move Fast and Break Things: How Facebook, Google, and Amazon Cornered Culture and Undermined Democracy* (New York: Little, Brown, 2017); Julie E. Cohen, "Law for the Platform Economy," *University of California Davis Law Review* 51 (2017): 133, 150. The power of Amazon to kick Parler off its platform reassured people horrified by the role of that social media in the violent assault on the Capitol on January 6, 2021—but exposed new worries about concentrated power.

101. Yochai Benkler, "Degrees of Freedom, Dimensions of Power," *Daedalus* 145 (2016): 18, 23.

102. See Jeffrey Rosen, "Richard S. Salant Lecture on Freedom of the Press," Harvard Kennedy School Shorenstein Center on Media, Politics and Public Policy, 2016, 14, 17, 23, 27, https://shorensteincenter.org/jeffrey-rosen-future-of-free-speech-in-a-digital-world/; Bryan X. Chen, *Always On: How the iPhone Unlocked the Anything-Anytime-Anywhere Future—and Locked Us In* (Cambridge, MA: Da Capo, 2012), 120.

103. See "'Fake News' Isn't Easy to Spot on Facebook, According to New Study," University of Texas News, Nov. 5, 2019, https://news.utexas.edu/2019/11/05/fake-news-isnt-easy-to-spot-on-facebook-according-to-new-study (citing Randall K. Minas and Alan R. Dennis, "Fake News on Social Media: People Believe What They Want to Believe When It Makes

No Sense at All," *Management Information Systems Quarterly*, Nov. 5, 2019, https://misq.org/misq/downloads/?___store=library).

104. Will Hellpern, "How 'Deceptive' Sponsored News Articles Could Be Tricking Readers—Even with a Disclosure Message," Business Insider, Mar. 17, 2016, https://www.businessinsider.com/how-deceptive-sponsored-news-articles-could-be-undermining-trusted-news-brands-even-with-a-disclosure-message-2016-3; Sue Shellenbarger, "Most Students Don't Know When News Is Fake, Stanford Study Finds," *Wall Street Journal*, Nov. 21, 2016, https://www.wsj.com/articles/most-students-dont-know-when-news-is-fake-stanford-study-finds-1479752576.

105. Jay Baer, "3 Ways to Fight Facebook's Algorithm and Customize Your Feed," Convince and Convert, Dec. 2015, http://www.convinceandconvert.com/socialmedia-tools/3-ways-to-fight-facebooks-algorithm-and-customize-your-feed.

106. Kate Van Huss, "What Is Algorithm in Digital Marketing and How Does It Affect Your Promotional Efforts," LinkedIn, Sept. 22, 2015, https://www.linkedin.com/pulse/what-algorithm-digital-marketing-how-does-affect-your-kate-van-huss.

107. Cathy O'Neil, *Weapons of Math Destruction: How Big Data Increases Inequality and Threatens Democracy* (New York: Broadway Books, 2016), 70–79.

108. "Fake News Expert on How False Stories Spread and Why People Believe Them," NPR, Dec. 14, 2016, https://www.npr.org/2016/12/14/505547295/fake-news-expert-on-how-false-stories-spread-and-why-people-believe-them.

109. See Seth Stephens-Davidowitz, *Everybody Lies: Big Data, New Data, and What the Internet Can Tell Us About Who We Really Are* (New York: Dey Street, 2017), 188–92.

110. Lincoln Caplan, "Should Facebook and Twitter Be Regulated Under the First Amendment?," Wired, Oct. 11, 2017, https://www.wired.com/story/should-facebook-and-twitter-be-regulated-under-the-first-amendment.

111. Rebecca Heilwewil, "How the 5G Coronavirus Conspiracy Went from Fringe to Mainstream," Vox, Apr. 24, 2020, https://www.vox.com/recode/2020/4/24/21231085/coronavirus-5g-conspiracy-theory-covid-facebook-youtube; Matthew Brown, "Fact Check: 5G Technology Is Not Linked to Coronavirus," *USA Today*, Apr. 23, 2020, https://www.usatoday.

com/story/news/factcheck/2020/04/23/fact-check-5-g-technology-not-linked-coronavirus/3006152001.

112. James Temperton, "How the 5G Coronavirus Conspiracy Theory Tore Through the Internet," Wired, Apr. 6, 2020, https://www.wired.co.uk/article/5g/coronavirus-conspiracy-theory; Abby Ohlheiser, "How Covid-19 Conspiracy Theorists Are Exploiting YouTube Culture," *MIT Technology Review*, May 7, 2020, https://wwwtechnologyreview.com/2020/05/07/1001252/youtube-covid-conspiracy-theories.

113. Katherine Schaeffer, "Nearly Three in Ten Americans Believe COVID-19 Was Made in a Lab," Pew Research Center, Apr. 8, 2020, https://www.pewresearch.org/fact-tank/2020/04/08/nearly-three-in-ten-americans-believe-covid-19-was-made-in-a-lab.

114. Sam Sabin, "Social Media Users Think Politicians Should Pay Heavy Price for Spreading False COVID-19 Information," Morning Consult, May 20, 2020, shttps://morningconsult.com/2020/05/20/coronavirus-politicians-misinformation-social-media-poll.

115. Hendrickson, "Critical in a Public Health Crisis"; Ella Koeze and Nathanial Popper, "The Virus Changed the Way We Internet," *New York Times*, Apr. 7, 2020, https://www.nytimes.com/interactive/2020/04/07/technology/coronavirus-Internet-use.html.

116. Paul Farhi and Sarah Ellison, "There's a Massive Audience for Coronavirus News Right Now. That Might Not Help the News Business," *Washington Post*, Mar. 17, 2020, https://www.washingtonpost.com/lifestyle/media/theres-a-massive-audience-for-coronavirus-news-right-now-that-might-not-help-the-news-business/2020/03/17/da89b06c-6861-11ea-9923-57073adce27c_story.html.

117. Mark Jurkowitz and Amy Mitchell, "Americans Who Primarily Get News Through Social Media Are Least Likely to Follow COVID-19 Coverage, Most Likely to Report Seeing Made-Up News," Pew Research Center, Mar. 25, 2020, https://www.journalism.org/2020/03/25/americans-who-primarily-get-news-through-social-media-are-least-likely-to-follow-covid-19-coverage-most-likely-to-report-seeing-made-up-news.

118. Sheera Frenkel, Davey Alba, and Raymond Zhong, "Surge of Virus Misinformation Stumps Facebook and Twitter," *New York Times*, Mar. 8, 2020, https://www.nytimes.com/2020/03/08/technology/coronavirus-misinformation-social-media.html.

119. Matt Richtel, "W.H.O. Fights a Pandemic Besides Coronavirus: An 'Infodemic,'" *New York Times*, Feb. 6, 2020, https://www.nytimes.com/2020/02/06/health/coronavirus-misinformation-social-media.html.

120. Victor Pickard, *Democracy Without Journalism? Confronting the Misinformation Society* (New York: Oxford University Press, 2019). As Fox News debunked President Trump's claims of reelection but for voter fraud in 2020, previously minor outlets dramatically gained viewers. See Michael M. Grynbaum and John Kblin, "Newsmax Rises Out of Obscurity After Refusing to Call Election," *New York Times*, Nov. 23, 2020.

Stepping in as local news sources vanish are websites, including more than a thousand that present "propaganda ordered up by dozens of conservative think tanks, political operatives, corporate executives and public-relations professionals." Davey Alba and Jack Nicas, "As Local News Dies, a Pay-for-Play Network Rises in Its Place," *New York Times*, Oct. 18. 2020, https://www.nytimes.com/2020/10/18/technology/timpone-local-news-metric-media

121. David Kaye, "Speech Police: The Global Struggle to Govern the Internet," Columbia Journalism Report, 2019, 14–21.

122. See John Wihbey, "Does Facebook Drive Political Polarization? Data Science and Research," *Journalist's Resource*, May 7, 2005, https://journalistsresource.org/studies/society/social-media/facebook-politicalpolarization-data-science-research. See also Levi Boxell et al., "Is the Internet Causing Political Polarization? Evidence from Demographics," Brown University, 2017, https://www.brown.edu/.Research/Shapiro/pdfs/age-polars.pdf. Polarization predates social media, and is also exacerbated by cable news with stark partisan division in their viewership. https://www.pewresearch.org/fact-tank/2020/04/01/americans-main-sources-for-political-news-vary-by-party-and-age/.

123. Samantha Power, "Why Foreign Propaganda Is More Dangerous Now," *New York Times*, Sept. 19, 2017, https://www.nytimes.com/2017/09/19/opinion/samantha-power-propaganda-fake-news.html. It is not just the Russians. See Gary King et al., "How the Chinese Government Fabricates Social Media Posts for Strategic Distraction, Not Engaged Argument," *American Political Science Review* 111 (2017): 484, https://gking.harvard.edu/files/gking/files/how_the_chinese_government_ fabricates_social_media_posts_for_strategic_distraction_not_engaged_argument.pdf.

124. Sam Meenasian, "The Biggest Threat Blowing Up Companies of All Sizes Worldwide," CNBC, Nov. 17, 2016, https://www.cnbc.com/

2016/11/17/cyber-hackers-the-biggest-threat-blowing-up-companies-worldwide.html.

125. Yochai Benkler, Robert Faris, Hal Robert, and Ethan Zuckerman, "Study: Breitbart-Led Right-Wing Media Ecosystem Altered Broader Media Agenda," *Columbia Journalism Review* (Mar. 3, 2017), https://www.cjr.org/analysis/breitbart-media-trumpharvard-study; Yochai Benkler, Robert Faris, Hal Roberts, and Ethan Zuckerman, "Study: Breitbart-Led Right-Wing Media Ecosystem Altered Broader Media Agenda," *Columbia Journalism Review*, Mar. 3, 2017, https://www.cjr.org/analysis/breitbart-media-trumpharvard-study.php; and Benkler, Faris, and Roberts, *Network Propaganda*.

126. Deepa Seetharaman et al., "Tone-Deaf: How Facebook Misread America's Mood on Russia," *Wall Street Journal*, Mar. 2, 2018, https://www.wsj.com/articles/tonedeaf-how-facebook-misread-americas-mood-on-russia-1520006034.

127. Edward Bernays, "Propaganda" (1928), History Is a Weapon, http://www.historyisaweapon.com/defcon1/bernprop.html, last visited Oct. 4, 2018. Propaganda tapping into rage seems particularly effective in the early part of the twenty-first century. See generally Peter Sloterdijk, *Rage and Time: A Psychopolitical Investigation*, trans. Mario Wenning (New York: Columbia University Press, 2006).

128. Phil Muncaster, "Bot-Driven Credential Stuffing Hits New Heights," *Infosecurity Magazine*, Feb. 21, 2018, https://www.infosecurity-magazine.com/news/botdriven-credential-stuffing-hits; Cristina Maza, "Florida Shooting: Russian Bots Flooded the Internet with Propaganda About Parkland Massacre," *Newsweek*, Feb. 16, 2018, http://www.newsweek.com/florida-shooting-russian-bots-twitter-809000.

129. Tom de Castella and Virginia Brown, "Trolling: Who Does It and Why?," BBC, Sept. 14, 2011, https://www.bbc.com/news/magazine-14898564; Howard Fosdick, "Why People Troll and How to Stop Them," OS News, Jan. 25, 2012, http://www.osnews.com/story/25540. Trolling and harassment on the internet often target political, racial, and sexual characteristics of individuals, and many more women identify problems then do men. Pew Research Center, "Online Harassment," July 11, 2017, https://www.pewresearch.org/Internet/2017/07/11/online-harassment-2017. Some suggest further that gendered interpretations of the First Amendment

shield too much troubling expression. See Mary Anne Franks, "Free Speech for the Last Girl," unpublished paper, 2020.

CHAPTER TWO

1. James Breid, "Early American Newspapering," *CW Journal*, Spring 2003, https://www.history.org/foundation/journal/spring03/journalism.cf.

2. This entire chapter benefits especially from the work of Richard D. Brown, *The Strength of a People: The Idea of an Informed Citizenry in America 1650–1870* (Chapel Hill: University of North Carolina Press, 1996); Christopher B. Daly, *Covering America: A Narrative History of a Nation's Journalism* (Amherst: University of Massachusetts Press, 2012); Sam Lebovic, *Free Speech and Unfree News: The Paradox of Press Freedom in America* (Cambridge, MA: Harvard University Press, 2016); Paul Starr, *The Creation of the Media: Political Origins of Modern Communications* (New York: Basic Books, 2004).

3. James Madison, "For the National Gazette," ca. December 19, 1791, Founders Online, June 13, 2018, https://founders.archives.gov/documents/Madison/01-14-02-0145.

4. See Colleen Sheehan, "The Politics of Public Opinion: James Madison's 'Notes on Government,'" *William and Mary Quarterly* 49 (1992): 609; Colleen Sheehan, *The Mind of James Madison* (New York: Cambridge University Press, 2017).

5. Brown, *The Strength of a People*, 24.

6. Virginia Declaration of Rights, section 12 (1776), U.S. Archives, https://www.archives.gov/founding-docs/virginia-declaration-of-rights (drafted by George Mason).

7. John Adams, *Diary and Autobiography of John Adams*, ed. L. H. Butterfield et al. (Cambridge, MA: Belknap Press, 1961), 120–121.

8. "A Summary of the 1765 Stamp Act," Colonial Williamsburg, http://www.history.org/history/teaching/tchcrsta.cfm, last visited Sept. 25, 2018; "Declaration of Rights and Grievances," October 14, 1774, Library of Congress, http://www.loc.gov/teachers/classroommaterials/presentationsandactivities/presentations/timeline/amrev/rebelln/rights.html, last visited Sept. 25, 2018.

9. Leonard W. Levy, *Emergence of a Free Press* (Chicago: Ivan R. Dee, 1985), 86–87.

10. Carol Sue Humphrey, *The American Revolution and the Press* (Evanston, IL: Northwestern University Press, 2013), 114–115.

11. Jerry W. Knudson, *Jefferson and the Press: Crucible of Liberty* (Columbia: University of South Carolina Press, 2006), 171 (quoting Thomas Jefferson). Jefferson later wrote a friend: "The basis of our governments being the opinion of the people, the very first object should be to keep that right; and were it left to me to decide whether we should have a government without newspapers or newspapers without a government, I should not hesitate a moment to prefer the latter" (Thomas Jefferson, "To Edward Carrington, Paris, Jan. 16, 1787," American History: From Revolution to Reconstruction and Beyond, http://www.let.rug.nl/usa/presidents/thomas-jefferson/letters-of-thomas-jefferson/jefl52.php, last visited Sept. 28, 2018.

12. Continental Congress, "An Appeal to the Inhabitants of Quebec" (1774), Digital History, University of Houston, http://www.digitalhistory.uh.edu/disp_textbook.cfm?smtID=3&psid=4104.

13. Levy, *Emergence of a Free Press*, xii.

14. Frederic Hudson, *Journalism in the United States, from 1690 to 1872* (New York: Harper and Brothers, 1873), 414.

15. Michael Warner, *The Letters of the Republic: Publication and the Public Sphere in Eighteenth-Century America* (Cambridge, MA: Harvard University Press, 1990), 19.

16. U.S. Constitution, Amendment I.

17. Lebovic, *Free Speech and Unfree News*; letter from Thomas Jefferson to Charles Yancy, Jan. 6, 1816, Founders Archive, https://founders.archives.gov/documents/Jefferson/03-09-02-0209, quoted by Justice White, *Miami Herald Publishing Co. v. Knight Newspapers*, 418 U.S. 241 259, 260 (1974).

18. Daly, *Covering America*, 30, 33, 41, 43, 50, 54.

19. Hudson, *Journalism in the United States*, 414.

20. Daly, *Covering America*, 54.

21. Thomas Jefferson to George Washington, quoted in Brown, *The Strength of a People*, 87; Jefferson's worries about falsehoods in the press are described on 88.

22. Mitchell Snay, *Horace Greeley and the Politics of Reform in Nineteenth-Century America* (Lanham, MD: Rowman and Littlefield, 2011), 50–51.

23. See David T. Z. Mindich, *Just the Facts: How "Objectivity" Came to Define American Journalism* (New York: New York University Press, 1998).

24. Henry Mayer, *All on Fire: William Lloyd Garrison and the Abolition of Slavery* (New York: W. W. Norton, 2008).

25. National Humanities Center, "Toolbox Library: The Black Press," http://nationalhumanitiescenter.org/pds/maai/community/text6/text6read.htm; Ida B. Wells, *Crusade for Justice: The Autobiography of Ida B. Wells*, ed. Alfreda M. Duster (Chicago: University of Chicago Press, 1970).

26. Peter G. Goheen, "The Impact of the Telegraph on the Newspaper in Mid-Nineteenth Century British North America," *Urban Geography* 11, no. 2 (1990): 107–129, DOI: 10.2747/0272-3638.11.2.107.

27. Frank Luther Mott, *A History of American Magazines*, vol. 3, *1865–1885* (Cambridge, MA: Harvard University Press, 1938); Frank Luther Mott, *A History of American Magazines*, vol. 4, *1885–1905* (Cambridge, MA: Harvard University Press, 1957).

28. Nancy Cott, *Fighting Words: The Bold American Journalists Who Brought the World Home Between the Wars* (New York: Basic Books, 2020), 11.

29. Richard R. John, *Spreading the News: The American Postal System from Franklin to Morse* (Cambridge, MA: Harvard University Press, 1992), 25, 191; Richard B. Kielbowicz, *News in the Mail: The Press, Post Office, and Public Information, 1700–1860s* (New York: Greenwood Press, 1989), 31–56, 141–155.

30. Tom Hazlett, "Ronald Coase and the Radio Spectrum," *Financial Times*, Dec. 15, 2009, https://www.ft.com/content/bfffd9fa-e9e2-11de-ae43-00144feab49a.

31. See *Grosjean v. American Press Co.*, 297 U.S. 233 (1936).

32. 297 U.S. 233, 250 (1936). See Richard C. Cortner, *The Kingfish and the Constitution: Huey Long, the First Amendment, and the Emergence of Modern Press Freedom in America* (Westport, CT: Greenwood Press, 1996).

33. *Grosjean v. American Press Co.*, 297 U.S. 233, 250 (1936).

34. See "Understanding Last Mile Internet Access," Medium, https://medium.com/@datapath_io/understanding-last-mile-internet-access-a62ee96c0a00, last visited Oct. 23, 2018.

35. "The Associated Press," Encyclopedia.com, https://www.encyclopedia.com/social-sciences-and-law/economics-business-and-labor/businesses-and-occupations/associated-press, last visited Oct. 24, 2018.

36. See generally Associated Press, "Our Story," https://www.ap.org/about/ourstory, last visited Sept. 26, 2018; "Network Effects: How a New Communications Technology Disrupted America's Newspaper Industry—in 1845," *The Economist*, Dec. 17, 2009, http://www.economist.com/node/15108618.

37. *Federal Communications Commission v. National Citizens Committee for Broadcasting*, 436 U.S. 775 (1978); C. Edwin Baker, *Media Concentration and Democracy: Why Ownership Matters* (Cambridge: Cambridge University Press, 2006). The FCC relaxed the cross-ownership rule in 2017, citing the need to assist media companies in competition with internet companies. Federal Communications Commission, "FCC Broadcast Ownership Rules," https://www.fcc.gov/consumers/guides/fccs-review-broadcast-ownership-rules. See C. Edwin Baker, "The Independent Significance of the Press Clause Under Existing Law," *Hofstra Law Review* 35 (2007): 955, 1010.

38. Robert McChesney and John Nichols, "The Rise of Professional Journalism: Reconsidering the Roots of Our Profession in an Age of Media Crisis," *In These Times*, Dec. 7. 2005, http://inthesetimes.com/article/2427/the_rise_of_professional_ journalism; Jack M. Balkin, "How to Regulate (and Not Regulate) Social Media," Knight Institute Occasional Paper Series, No. 1, Nov. 8, 2019, 5–6, https://papers.ssrn.com/sol3/papers.cfm?abstract_id=3484114.

39. Richard R. John, *Network Nation: Inventing American Telecommunications* (Cambridge, MA: Belknap Press, 2010), 8.

40. Gordon Gorier, "The Living Room Fixture," Radio Scribe, http://www.radioscribe.com/formats.html, last visited Oct. 24, 2018.

41. Hugh G. J. Aitkin, *The Continuous Wave: Technology and the American Radio, 1900–1932* (Princeton, NJ: Princeton University Press, 1985); Robert McChesney, *Telecommunications, Mass Media, and Democracy: The Battle for the Control of U.S. Broadcasting, 1928–1935* (New York: Oxford University Press, 1995), 12–18.

42. John, *Network Nation*, 8.

43. See Gerald J. Baldasty, *The Commercialization of News in the Nineteenth Century* (Madison: University of Wisconsin Press, 1992), 28; Michael Schudson, "The Objectivity Norm in American Journalism," *Journalism* 2 (2001): 149, 159–160; Jack Shafer, "The Lost World of Joseph Pulitzer," *Slate*, Sept. 16, 2005, http://www.slate.com/articles/news_and_

politics/press_box/2005/09/the_lost_world_of_joseph_pulitzer.html; see generally Géraldine Muhlmann, *A Political History of Journalism*, trans. Jean Birrell (Cambridge: Polity Press, 2008).

44. Phyllis Leslie Abramson, *Sob Sister Journalism* (New York: Greenwood Press, 1990); David R. Spencer, *The Yellow Journalism: The Press and America's Emergence as a World Power* (Evanston, IL: Northwestern University Press, 2007).

45. "U.S. Diplomacy and Yellow Journalism, 1895–1898," Office of the Historian, Department of State, https://history.state.gov/milestones/1866-1898/yellow-journalism, last visited Dec. 18, 2018; see generally Michael Schudson, *Discovering the News: A Social History of American Newspapers* (New York: Basic Books, 1978); Spencer, *The Yellow Journalism*.

46. See generally Fred W. Friendly, *Minnesota Rag: The Dramatic Story of the Landmark Supreme Court Case That Gave New Meaning to the Freedom of the Press* (New York: Random House, 1981).

47. See James L. Baughman, "The Fall and Rise of Partisan Journalism," Center for Journalism Ethics, University of Wisconsin, Apr. 20, 2011, https://ethics.journalism.wisc.edu/2011/04/20/the-fall-and-rise-of-partisan-journalism.

48. David Goodman, *Radio's Civic Ambition: American Broadcasting and Democracy in the 1930s* (New York: Oxford University Press, 2011).

49. "Radio: The Golden Age Around the World," Britannica, https://www.britannica.com/topic/radio/The-Golden-Age-around-theworld, last visited Oct. 3, 2018.

50. Stuart N. Brotman, "Revisiting the Public Interest Standard in Communications Law and Regulation," Brookings Institution, Mar. 23, 2017, https://www.brookings.edu/research/revisiting-the-broadcast-public-interest-standard-in-communications-law-and-regulation.

51. Tona Hangen, "When Radio Rules: The Social Life of Sound," *American Quarterly* 66 (June 2014): 465.

52. Steve Pociask, "A New Look at Media Cross-Ownership Rules," *Forbes*, Nov. 7, 2017, https://www.forbes.com/sites/stevepociask/2017/11/07/a-new-look-atmedia-cross-ownership-rules/#55854a24181f.

53. See Gerd Horten, *Radio Goes to War: The Cultural Politics of Propaganda During World War II* (Berkeley: University of California Press, 2002); R. Jason Loviglio, *Radio's Intimate Public: Network Broadcasting and Mass-Mediated Democracy* 1–37 (Minneapolis: University of Minnesota

Press, 2005); Judith E. Smith, "Literary Radicals in Radio's Public Sphere," unpublished paper, 2019.

54. See Thomas W. Hazlett, David Porter, and Vernon Smith, "Radio Spectrum and the Disruptive Clarity of Ronald Coase," *Journal of Law and Economics* 54, no. 4 (2011): S125–165, DOI: 10.1086/662992.

55. Charles L. Ponce de Leon, *That's the Way It Is: A History of Television News in America* (Chicago: University of Chicago Press, 2015).

56. Erik Barnouw, *Tube of Plenty: Evolution of American Television*, 2nd ed. (New York: Oxford University Press, 1990), 549–560.

57. See Susan Douglas, *Listening In: Radio and the American Imagination* (Minneapolis: University of Minnesota Press, 1999), 161–198.

58. *FCC v. League of Women Voters*, 468 U.S. 364 (1984); "The History of Public Broadcasting," University of North Carolina School of Government, Oct. 21, 2013, https://onlinempa.unc.edu/history-of-public-broadcasting.

59. President John F. Kennedy, "Statement by the President upon Signing Bill Providing for Educational Television," May 1, 1962, American Presidency Project, https://www.presidency.ucsb.edu/documents/statement-the-president-upon-signing-bill-providing-for-educational-television.

60. WGBH, "Time-Line: The History of Public Broadcasting in the U.S." (describing the 1962 All Channels Receiver Act, 47 U.S.C. § 303(s)). As Chairman of the Federal Communications commission during this period, Newton Minow shepherded this statute, the first communications satellite, and the framework for a national public media.

61. For a brief history of federal aid to public media in the United States, see Lee Banville, "The Story of US Public Media," SwissInfo, Feb. 6, 2018, https://www.swissinfo.ch/eng/npr--pbs-and-cpb_the-story-of-us-public-media/43877508. See also Glenn J. McLoughlin and Lena A. Gomez, "The Corporation for Public Broadcasting: Federal Funding and Issues," Congressional Research Service, May 3, 2017, https://fas.org/sgp/crs/misc/RS22168.pdf.

62. National Cable and Telecommunications Association, "History of Cable Television," 2010, https://web.archive.org/web/20100905133543/http://www.ncta.com/About/About/HistoryofCableTelevision.aspx.

63. Adrian E. Herbst, Gary R. Matz, and John F. Gibbs, "A Review of Federal, State and Local Regulation of Cable Television in the United States," *William Mitchell Law Review* 10 (1984): 377, http://open.mitchellhamline.edu/wmlr/vol10/iss3/1.

64. See "Media Bias/Fact Check," MSNBC, https://mediabiasfactcheck.com/msnbc, last visited Dec. 18, 2018.

65. See Carter Goodrich, *Government Promotion of American Canals and Railroads 1800–1890* (New York: Columbia University Press, 1960); James Willard Hurst, *Law and Economic Growth: The Legal History of the Lumber Industry in Wisconsin, 1836–1915* (Madison: University of Wisconsin Press, 1984).

66. Jeff Madrick, "Innovation: The Government Was Crucial After All," *New York Review of Books*, Apr. 14, 2014, http://www.nybooks.com/articles/2014/04/24/innovation-government-wascrucial-after-all (reviewing Mariana Mazzucato, *The Entrepreneurial State: Debunking Public vs. Private Sector Myths* [New York: Public Affairs, 2011] and William H. Janeway, *Doing Capitalism in the Innovation Economy: Markets, Speculation and the State* [Cambridge: Cambridge University Press, 2012]).

67. "ARPANET," Defense Advanced Research Projects Agency, https://www.darpa.mil/about-us/timeline/arpanet, last visited Oct. 4, 2018; Madrick, "Innovation: The Government Was Crucial After All."

68. Ethan Baron, "Google, Tesla, Apple, Facebook Rake in Massive Subsidies: Report," *Mercury News*, July 3, 2018, https://www.mercurynews.com/2018/07/03/google-tesla-apple-facebook-rake-in-massive-subsidies-report.

69. Timothy Lee, "Network Neutrality, Explained," Vox, May 21, 2015, https://www.vox.com/cards/network-neutrality/what-have-federal-regulators-done-toprotect-network-neutrality.

70. See "CDA 230: Legislative History," Electronic Freedom Foundation, https://www.eff.org/issues/cda230/legislative-history, last visited Dec. 31, 2018.

71. See *Mozilla Corp. v. FCC*, 940 F.3d 1 (D.C. Cir. 2019); Keith Collins, "Net Neutrality Has Officially Been Repealed," *New York Times*, June 11, 2018, https://www.nytimes.com/2018/06/11/technology/net-neutrality-repeal.html; Cecilia Kang, "Justice Department Sues to Stop California Net Neutrality Law," *New York Times*, Sept. 30, 2018, https://www.nytimes.com/2018/09/30/technology/net-neutrality-california.html.

72. Section 230 says that "No provider or user of an interactive computer service shall be treated as the publisher or speaker of any information provided by another information content provider" (47 U.S.C. §230). Accordingly, internet intermediaries that host or republish speech are protected against laws that might otherwise make them legally—and financially—responsible for speech they carry and distribute.

73. See generally Douglas K. Smith et al., *Table Stakes: A Manual for Getting in the Game of News* (Monee, IL: Knight Foundation, 2017).

74. See generally Jonathan Mandell, "8 Ways Television Is Influencing Theater," HowlRound, Oct. 16, 2013, https://howlround.com/8-ways-television-influencingtheater; Tom Lewis, "'A Godlike Presence': The Impact of Radio on the 1920s and 1930s," *OAH Magazine of History*, Spring 1992, 26.

75. Matt Pressberg, "Amazon, Netflix Should Consider Buying Movie Theaters," The Information, Jan. 16, 2018, https://www.theinformation.com/articles/amazon-netflix-should-consider-buying-movie-theaters.

76. "Digital Music: Float of a Celestial Jukebox," *The Economist*, Jan. 11, 2018, https://www.economist.com/business/2018/01/11/having-rescued-recorded-music-spotify-may-upend-the-industry-again.

77. C. W. Anderson, Leonard Downie Jr., and Michael Schudson, *The News Media: What Everyone Needs to Know* (New York: Oxford University Press, 2016), 120–121.

78. Sebastian Anthony, "Is the Internet a Failed Utopia? Is the Internet Still an Equal Playing Field, or Are the Megacorps Taking Over?," Ars Technica, June 13, 2015, https://arstechnica.com/information-technology/2015/06/is-theinternet-a-failed-utopia.

79. Jack Balkin, "Old-School/New-School Speech Regulation," *Harvard Law Review* 127 (2014): 2296; Margaret E. Roberts, *Censored: Distraction and Diversion Inside China's Great Firewall* (Princeton, NJ: Princeton University Press, 2018), 41–92; Zeynup Tufekci, *Twitter and Tear Gas: The Power and Fragility of Networked Protest* (New Haven, CT: Yale University Press, 2017), 239.

80. Jill Abramson, *Merchants of Truth* (New York: Simon and Schuster, 2019); Ann Marie Lipinski, "Relating Journalism's Age of Anxiety: Review of Jill Abramson, 'Merchants of Truth,'" *Chicago Tribune*, Feb. 10, 2019.

81. *American Broadcasting Co. v. Areo*, 573 U.S. 431 (2014).

82. Kristine A. Oswold, "Mass Media and the Transformation of American Politics," *Marquette Law Rev*iew 77 (2009): 395.

83. Robert E. Litan, "Antitrust Enforcement and the Telecommunications Revolution: Friends, Not Enemies," speech at the National Academy of Engineering, Washington, DC, Oct. 6, 1994, https://www.justice.gov/atr/speech/antitrust-enforcement-and-telecommunications-revolution.

84. See Steve Coll, *The Deal of the Century: The Breakup of AT&T* (New York: Open Road Media, 1986).

85. See Christopher H. Sterling, Phyllis W. Bernt, and Martin B. H. Weiss, *Shaping American Telecommunications: A History of Technology, Policy, and Economics* (Mahwah, NJ: Lawrence Erlbaum Associates, 2005), 270–279; Cecilia Kang, "TV's Future: FCC Decisions on Internet Access, Comcast-NBC Merger Approaching," *Washington Post*, Dec. 18, 2010, http://www.washingtonpost.com/wp-dyn/content/article/2010/12/17/AR2010121703792.html; Kristina M. Lagasse, "Shaping the Future of the Internet: Regulating the World's Most Powerful Information Resource in *U.S. Telecom Ass'n v. FCC*," *Loyola Law Review* 63 (2017): 322, 323.

86. See Victor Pickard, *Democracy Without Journalism? Confronting the Misinformation Society* (New York: Oxford University Press, 2019).

87. 15 U.S.C. § 1802 (2012).

88. Maurice E. Stucke and Allen P. Grunes, "Why More Antitrust Immunity for the Media Is a Bad Idea," *Northwestern University Law Review Colloquy* 105 (2010): 115, 127. See Jason A. Martin, "Reversing the Erosion of Editorial Diversity: How the Newspaper Preservation Act Has Failed and What Can Be Done," *Communications Law and Policy* 13 (2008): 63.

89. See Jack Goldsmith and Tim Wu, *Who Controls the Internet? Illusions of a Borderless World* (New York: Oxford University Press, 2008); Kirsten Grind, Sam Schechner, Robert McMillan, and John West, "How Google Interferes with Its Search Algorithms and Changes Your Results," *Wall Street Journal*, Nov. 15, 2019, https:www.wsj.com/articles/how-google-interferes-with-its-search-algorithms-and-changes-your-results-11573823753.

90. See McChesney, *Telecommunications, Mass Media, and Democracy*; Eugene E. Leach, "Tuning Out Education, Chapter 1," *Current*, Jan. 14, 1983, https://current.org/1983/01/tuning-out-education.

91. Virginia Declaration of Rights, Library of Congress, https://www.loc.gov/resource/rbpe.17802200/?sp=2, last visited Oct. 3, 2018.

CHAPTER THREE

1. Victor Pickard, *Democracy Without Journalism? Confronting the Misinformation Society* (New York: Oxford University Press, 2019).

2. Compare *Citizens United v. Federal Election Commission*, 558 U.S. 310 (2010), with *Red Lion Broadcasting v. FCC*, 395 U.S. 367 (1969).

3. See *Barr v. American Association of Political Consultants*, See Barr v. American Assn. Polit. Consultants, ___ U.S. ___ (2020) (Court upholds general ban on robo calls while striking down content-based exception).

4. See *Gary B. v. Whitman*, _____ F.3rd _____ (en banc dismissed) (settlement after district court recognized federal right to education).

5. Tim Wu, "Is the First Amendment Obsolete?," *Michigan Law Review* 117 (2018): 547.

6. Sam Lebovic, *Free Speech and Unfree News: The Paradox of Press Freedom in America* (Cambridge, MA: Harvard University Press, 2016), 2–10, 32–44. See Christopher B. Daly, *Covering America: A Narrative History of a Nation's Journalism* (Amherst: University of Massachusetts Press, 2012), 46–58.

7. Letter from Thomas Jefferson to Charles Yancy, Jan. 6, 1816, Founders Archive, https://founders.archives.gov/documents/Jefferson/03-09-02-0209, quoted by Justice White, *Miami Herald Publishing Co. v. Knight Newspapers*, 418 U.S. 241 259, 260 (1974).

8. Over time, the First Amendment has applied also to actions of courts and of lower executive officials—but not yet explicitly to the president of the United States. See Sonya West, "Suing the President for First Amendment Violations," *Oklahoma Law Review* 71 (2018): 321, https://digitalcommons.law.ou.edu/cgi/viewcontent.cgi?article=1349&context=olr.

9. *Barron v. Baltimore*, 32 U.S. (7 Pet.) 243 (1833).

10. Leonard W. Levy, *Origins of the Bill of Rights* (New Haven, CT: Yale University Press, 1999), 119.

11. Leonard W. Levy, *Emergence of a Free Press* (Chicago: Ivan R. Dee, 1985), x–xi, xv–xvi.

12. *New York Times v. Sullivan*, 376 U.S. 354 (1964); *Curtis Publishing v. Butts*, 388 U.S. 130 (1967); *Gertz v. Robert Welch, Inc.*, 418 U.S. 323 (1974).

13. See Joseph Story, *Commentaries on the Constitution*, vol. 3 (Boston: Gray, 1833), sec. 1874–1886, reprinted in Philip Kurland, ed., *The Founders' Constitution* (1987), 4:182–185 (describing constitutional protection for speaking, writing, and printing opinions without prior restraint so long as the expression does not injure another person, disturb the peace, try to subvert the government, or is otherwise illegal).

14. Library of Congress Research Guides, "Alien and Sedition Acts: Primary Documents in American History," https://guides.loc.gov/alien-and-sedition-acts.

15. See *Gitlow v. New York,* 268 U.S. 652 (1925).

16. Geoffrey Stone, *Perilous Times: Free Speech in Wartime* (New York: W. W. Norton, 2004).

17. Laura Weinrib, *The Taming of Free Speech: America's Civil Liberties Compromise* (Cambridge, MA: Harvard University Press, 2016), 11–12, 33–36, 41–44, 70–80, 189, 202–203.

18. See *Abrams v. United States*, 250 U.S. 616, 624 (1919) (Holmes, J., dissenting); *Debs v. United States*, 249 U.S. 211 (1919); *Gitlow v. New York*, 268 U.S. 652, 672 (1925) (Holmes, J., dissenting); *Whitney v. California*, 274 U.S. 257, 372 (1927) (Brandeis, J., concurring). See David L. Hudson Jr., "How 2 Supreme Court Cases from 1919 Shaped the Next Century of First Amendment Law," *ABA Journal*, March 12, 2019, https://www.abajournal.com/web/article/how-two-supreme-court-cases-from-1919-shaped-the-next-century-of-first-amendment-law; Laura Weinrib, "The Limits of Dissent: Reassessing the Legacy of the World War I Free Speech Cases," *Journal of Supreme Court History* 44 (2019): 278.

19. Thomas Healy, *The Great Dissent: How Oliver Wendell Holmes Changed His Mind—and Changed the History of Free Speech in America* (New York: Henry Holt, 2013).

20. Philippa Strum, *Louis D. Brandeis: Justice for the People* (New York: Schocken Books, 1984), 138, 404, 406, 414–416.

21. Gerald Berk, *Louis D. Brandeis and the Making of Regulated Competition, 1900–1932* (Cambridge: Cambridge University Press, 2009), 44.

22. Robert F. Nagel, "How Useful Is Judicial Review in Free Speech Cases," *Cornell Law Review* 69 (1984): 302.

23. Stone, *Perilous Times.*

24. See David A. Anderson, "The Origins of the Press Clause," *UCLA Law Review* 30 (1983): 456–460; RonNell Andersen Jones and Sonja R. West, *The Fragility of the Free American Press*, 112 NW. U. L. REV. 48, 52–54 (2017); Potter Stewart, "Or of the Press," *Hastings Law Journal* 26 (1975): 631.

25. *N.Y. Times Co. v. Sullivan*, 376 U.S. 254, 278–80 (1964).

26. See *Red Lion Broadcasting Co. v. FCC*, 395 U.S. 367, 369, 400–401 (1969).

27. Communications Act of 1934, 47 U.S.C. § 151 (amended to cover television and newer technologies); Stuart N. Brotman, "Revising the Broadcast Public Interest Standard in Communications Law and Regulation," Brookings Institution, Mar. 23, 2017, https://www.brookings.edu/research/revisiting-the-broadcast-public-interest-standard-in-communications-law-and-regulation; *Red Lion Broadcasting Co. v. FCC*, 395 U.S. 367 (1969). Congress borrowed the phrase "public interest, convenience, and necessity" from the Interstate Commerce Commission's authority, analogizing airwaves to roadways.

28. David Goodman, *Radio's Civic Ambition: American Broadcasting and Democracy in the 1930s* (New York: Oxford University Press, 2011); Tona Hangen, "When Radio Ruled: The Social Life of Sound," *American Quarterly* 66 (June 2014): 465.

29. *Red Lion Broadcasting Co. v. FCC*, 395 U.S. 367 (1969); *Turner Broadcasting System v. FCC*, 512 U.S. 622 (1994); *Reno v. ACLU*, 521 U.S. 844 (1997). See also *FCC v. Pacifica Foundation*, 438 U.S. 726 (1978). And see Glen O. Robinson, "The Electronic First Amendment: An Essay for the New Age," *Duke Law Journal* 47 (1998): 899, 903; Jennifer L. Polse, "*United States v. Playboy Entertainment Group, Inc.*," *Berkeley Technology Law Journal* 16 (2001): 347, 348; Josephine Soriano, "The Digital Transition and the First Amendment: Is It Time to Reevaluate *Red Lion*'s Scarcity Rationale?," *Boston University Public Interest Law Journal* 15 (2006): 341.

30. *Red Lion Broadcasting Co. v. FCC*, 395 U.S. 367, 400 (1969).

31. *CBS, Inc., v. FCC*, 453 U.S. 367 (1981); *Columbia Broadcasting, Inc. v. Democratic National Committee*, 412 U.S. 94 (1973). And a public broadcaster may decide to select and exclude candidates for a debate much as a public university may choose whom to select as speakers. *Arkansas Educational Television Comm'n v. Forbes*, 523 U.S. 666 (1998).

32. C. Edwin Baker, "The Independent Significance of the Press Clause Under Existing Law," *Hofstra Law Review* 35 (2007): 955, 980–982; C. Edwin Baker, "*Turner Broadcasting*: Content-Based Regulation of Persons or Presses," *Supreme Court Review*, 1994, 57–111.

33. Dante Chinni, "Is the Fairness Doctrine Fair Game?," Pew Research Center, July 19, 2007, http://www.pewresearch.org/2007/07/19/is-the-fairness-doctrine-fair-game. Note that rather than relying on scarcity for media regulation, Canada predicates its governmental oversight of media on cultural commitments.

34. 395 U.S. 367, 390 (1969).

35. Richard D. Brown, *The Strength of a People: The Idea of an Informed Citizenry in America, 1650–1870* (Chapel Hill: University of North Carolina Press, 1996).

36. 395 U.S. 367, 390 (1969).

37. Baker, "*Turner Broadcasting*: Content-Based Regulation of Persons or Presses," 101.

38. *Packingham v. North Carolina*, 582 U.S. ___ (2017) (rejecting state law criminalizing access by a registered sex offender to a commercial social networking website).

39. Daly, *Covering America*, 25–26 (2012) (citing Benjamin Franklin).

40. Daly, *Covering America*, 290 (2012).

41. Daly, *Covering America*, 299 (quoting Murrow). When Senator McCarthy prepared to attack Murrow himself, Murrow devoted an episode of his show to a sustained examination and attack on McCarthy and in defense of free speech and dissent. Broadcast company CBS and the show's sponsor, Alcoa, initially stood with Murrow, but after a while the network reduced Murrow's prominence and the sponsor halted funding for his show. Later pressures on the independence, diversity, and fairness of news reporting emerged due to concentrated ownership and profit-seeking by big media companies (396–397).

42. Brooks Boliek, "FCC Finally Kills Off Fairness Doctrine," Politico, Aug. 22, 2011.

43. See Gabriel Sherman, *The Loudest Voice in the Room: How the Brilliant, Bombastic Roger Ailes Built Fox News—and Divided a Country* (New York: Random House, 2017).

44. Kevin M. Cruse and Julian Zelizer, "How Policy Decisions Spawned Today's Hyperpolarized Media," *Washington Post*, Jan. 17, 2019, https://www.washingtonpost.com/outlook/2019/01/17/how-policy-decisions-spawned-todays-hyperpolarized-media. For Roger Ailes's strategy, see Sherman, *The Loudest Voice in the Room*; for his philosophy, see Roger Ailes with Jon Kraushar, *You Are the Message: Getting What You Want by Being Who You Are* (New York: Crown Business, 1988).

45. Jane Mayer, "The Making of the Fox News White House," *New Yorker*, March 11, 2019, https://www.newyorker.com/magazine/2019/03/11/the-making-of-the-fox-news-white-house.

46. John Gramlich, "Five Facts About Fox News," Pew Research Center, April 8, 2020, https://www.pewresearch.org/fact-tank/2020/04/08/five-facts-about-fox-news. See also Sherman, *The Loudest Voice in the Room*; Tim Dickinson, "How Roger Ailes Built the Fox News Fear Factory," *Rolling Stone*, June 9, 2011, https://www.rollingstone.com/politics/politics-news/how-roger-ailes-built-the-fox-news-fear-factory-244652 (tracing ties between Fox and the Republican Party); William Falk, "Why Fox News Was Created," *The Week*, Nov. 22, 2019, https://theweek.com/articles/880107/why-fox-news-created (same).

47. Kalev Leetaru, "Why Was 2017 the Year of the Filter Bubble?," *Forbes*, Dec. 18, 2017, https://www.forbes.com/sites/kalevleetaru/2017/12/18/why-was-2017-the-year-of-the-filter-bubble/#5effbf88746b.

48. Christopher A. Bail et al., "Exposure to Opposing Views on Social Media Can Increase Political Polarization," *Proceedings of the National Academy of Sciences* 115 (2018): 9216; Matthew Gentzkow and Jesse M. Shapiro, "Ideological Segregation Online and Offline," *Quarterly Journal of Economics* 126 (2011): 1802. On audience share, see Aaron Pressman, "For the First Time, More Americans Pay For Internet Video Than Cable or Satellite TV," *Fortune*, Mar. 19, 2019, https://fortune.com/2019/03/19/cord-cutting-record-netflix-deloitte/; Right Idea: Media and Creative, Television Adveritising: Cable vs Broadcast, Mar. 26, 2018, http://info.rightideacreative.com/television-advertising-cable-vs-broadcast.

49. *Denver Area Educational Telecommunications Consortium v. FCC*, 518 U.S. 727 (1996) (Breyer, J., plurality opinion).

50. See *Ashcroft v. American Civil Liberties Union*, 535 U.S. 564 (2002) (Ashcroft I); *Ashcroft v. American Civil Liberties Union II*, 542 U.S. 656 (2004) (Ashcroft II); *Am. Civil Liberties Union v. Gonzales*, 478 F. Supp. 2d 775 (E.D. Pa. 2007), aff'd *Am. Civil Liberties Union v. Mukasey*, 534 F.3d 181 (3d Cir. 2008), cert. denied, *Mukasey v. Am. Civil Liberties Union*, 555 U.S. 1137 (2009).

51. *Denver Area Educ. Telecomm. Consortium, Inc. v. FCC*, 518 U.S. 727 (1996).

52. FCC, "Consumer Guide: Obscene, Indecent and Profane Broadcasts," 2017, https://transition.fcc.gov/cgb/consumerfacts/obscene.pdf.

53. *Turner Broadcasting System v. FCC*, 520 U.S. 180, 194 (1997).

54. See "Applicability of Sponsor Identification Rules," 40 FCC 141 (1963), modified 40 *Fed. Reg.* 41926 (Sept. 9, 1975). Since the Radio Act

of 1927, Congress has required broadcasters to identify their sponsors. In 1969, the FCC applied similar disclosure rules to cable operators for programming they create. Richard Kielbowicz and Linda Lawson, "Unmasking Hidden Commercials in Broadcasting: Origins of the Sponsorship Identification Regulations, 1927–1963," *Federal Communication Law Journal* 56 (2004): 331; Steven Waldman, "The Information Needs of Communities: The Changing Media Landscape in a Broadband Age," Federal Communications Commission Working Group on Information Needs of Communities, 2011, 278. See Rita Marie Cain Reid, "Disclosure and Deception: Regulation of Material Connections Between Ad Sponsors and Their 'Endorsers' in New and Traditional Media," Aug. 2011, https://papers.ssrn.com/sol3/papers.cfm?abstract_id=1905137. Changes in statutory interpretation and regulations—changes reflecting political choices—have made required disclosures less than effective, but have not (at least not yet) been accompanied by a Supreme Court rejection of governmental power to require disclosure of funded material. Meredith McGehee, "Listeners Are Entitled to Know by Whom They Are Being Persuaded," Huffington Post, Jan. 14, 2014, https://www.huffpost.com/entry/listeners-are-entitled-political-advertising_b_4591415.

55. Federal Communications Commission, "Consumer Guides: Hoaxes," https://www.fcc.gov/reports-research/guides/hoaxes (explaining 47 C.F.R. § 73.1217).

56. Compare 567 U.S. 709, 7333 (Breyer, J., concurring in the judgment) (seeing the value of false statements to prevent embarrassment or stop a panic); *United States v. Alvarez, Abrams v. United States*, 250 U.S. 616, 630 (1919) (Holmes, J., dissenting) (valuing falsities within the free trade of ideas) with *Hustler Magazine, Inc. v. Falwell*, 485 U.S. 46, 56 (1988); *Gertz v. Robert Welch, Inc.*, 418 U.S. 323, 340–41 (1974),

57. See *United States v. Playboy Entm't Grp., Inc.*, 529 U.S. 803 (2000); *Sable Commc'ns of Cal., Inc. v. FCC*, 492 U.S. 115 (1989).

58. Baker, "The Independent Significance of the Press Clause Under Existing Law," 980–984.

59. James Grimmelmann, "Listeners' Choices," *Colorado Law Review* 90, no. 2 (2019): 377–379, 400–401 (citing cases).

60. *Abood v. Detroit Bd. of Ed.*, 431 U.S. 209 (1977).

61. *Connick v. Meyers*, 461 U.S. 138 (1983); *Garcetti v. Ceballos*, 543 U.S. 1186 (2006).

62. See *Citizens United v. FEC*, 558 U.S. 310 (2010).

63. *Janus v. AFSCME*, 138 S. Ct. 2448, 2501 (2018) (Kagan, J., dissenting).

64. *NIFLA v. Becerra*, 138 S. Ct. 2361, (2018) (Breyer, J., dissenting).

65. 198 U.S. 45 (1905).

66. *West Coast Hotel v. Parrish*, 300 U.S. 379 (1937).

67. Jedidiah Purdy, "Beyond the Bosses' Constitution: The First Amendment and Class Entrenchment," *Columbia Law Review* 118 (2018): 2161. Amy Kapczynski, "The Lochnerized First Amendment and the FDA: Toward a More Democratic Political Economy," *Columbia Law Review Forum*, 2018, https://columbialawreview.org/content/the-lochnerized-first-amendment-and-the-fda-toward-a-more-democratic-political-economy; Robert Post and Amanda Shanor, "Adam Smith's First Amendment," *Harvard Law Review Forum* 128 (2015): 165; but see Jane R. Bambauer and Derek E. Bambauer, "Information Libertarianism," *California Law Review* 105 (2017): 339.

68. 138 S. Ct. 2361 (2018), 585 U.S. ___ (2018).

69. See Jeremy Kessler, "The Early Years of First Amendment Lochnerism," *Columbia Law Review* 116 (2016): 1915, https://columbialawreview.org/wp-content/uploads/2016/12/Kessler-J.pdf.

70. *Douglas v. City of Jeanette*, 319 U.S. 157, 179, 181–82 (Jackson, J., concurring in part and dissenting in part).

71. See *National Institute of Family and Life Advocates v. Becerra*, 585 U.S. ___, __ (2018).

72. Weinrib, *The Taming of Free Speech*, 319.

73. Kessler, "The Early Years of First Amendment Lochnerism," 1925.

74. The phrase has been attributed to President Abraham Lincoln, used in a dissenting opinion by Justice Robert Jackson, and associated with reflections by Thomas Jefferson. *Terminiello v. City of Chicago*, 3337 U.S. 1, 37 (1949) (Jackson, J., dissenting).

75. See *Bigelow v. Virginia*, 421 U.S. 8091 (1975); *Virginia State Board of Pharmacy v. Virginia Citizens Consumer Council, Inc.*, 425 U.S. 748 (1976). See also *Central Hudson Gas & Electric Corp. v. Public Service Commission of New York*, 447 U.S. 557, 566 (1980).

76. See *Loan Payment Admin v. Hubanks*, No. 14-CV-04420-LHK, 2018 WL 6438362, at *7 (N.D. Cal. Dec. 7, 2018); Reply Brief for Plaintiff-Appellant at 19, *Exxon Mobil v. Healey*, 17-CV-2301, 2018 WL 5298251 (C.A. 2), at *19.

77. John C. Coates IV, Corporate Speech & The First Amendment: History, Data, and Implications, 30 Const. Comment. 223, 224 (2015).

78. *Zauderer v. Office of Disciplinary Counsel*, 471 U.S. 626, 651 (1985).

79. *Citizens United v. FEC*, 558 U.S. 310 (2010); Reporters Committee for Freedom of the Press, "Newsflash! *Citizens United* Has Been Good for Campaign Finance Transparency," 2012, https://www.rcfp.org/journals/news-media-and-law-winter-2012/newsflash-citizens-united-h; David L. Hudson Jr., "4th Circuit Invalidates Maryland Disclosure Law on Internet Political Ads," Free Speech Center, Middle Tennessee State University, Technology and Marketing Law Blog, Dec. 12, 2019, https://www.mtsu.edu/first-amendment/post/412/4th-circuit-invalidates-maryland-disclosure-law-on-internet-political-ads.

80. See 47 C.F.R. § 73.1217.

81. *Washington Post v. McManus*, No. 19-1132 (4th Cir. 2019); see *American Meat Institute v. USDA*, 760 F.3d 718 (D.C. Cir. 2014) (en banc).

82. Jennifer M. Keighley, "Can You Handle the Truth? Compelled Commercial Speech and the First Amendment," *Journal of Constitutional Law* 15 (2012): 541; Eugene Volokh, "The Law of Compelled Speech," *Texas Law Review* 87 (2018): 355.

83. Eric Goldman, "Eric's Comments" (comment on Venkat Balasubramani, "Maryland Disclosure Requirements for Online Political Ads Violates the First Amendment—Washington Post v. McManus"), Technology and Marketing Law Blog, December 18, 2019, https://blog.ericgoldman.org/archives/2019/12/maryland-disclosure-requirements-for-online-political-ads-violates-the-first-amendment-washington-post-v-mcmanus.htm.

84. "Note: A Shield for David and a Sword Against Goliath: Protecting Association While Combatting Dark Money Through Proportionality," 133 *Harvard Law Review* 643 (2019).

85. See Valerie C. Brannon, "Regulating Big Tech: Legal Implications," Version 7, Congressional Research Service, LSB10309, Sept. 11, 2019, 3–4.

86. *Chaplinsky v. New Hampshire*, 315 U.S. 568, 571–72 (1942), narrowed by *Terminiello v. Chicago*, 337 U.S. 1 (1949).

87. *New York Times Co. v. Sullivan*, 376 U.S. 254 (1964); *Curtis Publishing Co. and Associated Press v. Walker*, 388 U.S. 130 (1967).

88. *Rosenblatt v. Baer*, 383 U.S. 75 (1966) (remanding to allow the trial court to decide if the target was a public official).

89. *United States v. Alvarez*, 567 U.S. 709 (2012).

90. *Beauharnais v. Illinois*, 343 U.S. 250 (1956). See also Rohan Pavularui, "*Sapiro vs. Ford*: The Mastermind of the Marshall Maneuver," Exposé Magazine (Harvard University), 2016, https://projects.iq.harvard.edu/expose/pavuluri (high-profile settlement of group libel case included Henry Ford's recantation of anti-Semitic statements).

91. See Jeremy Waldron, *The Harm in Hate Speech* (Cambridge, MA: Harvard University Press, 2012).

92. See *Neiman-Marcus v. Lait*, 13 F.R.D. 311 (S.D.N.Y. 1952); *R.A.V. v. City of St. Paul*, 505 U.S. 377, 383 (1992). But see *Virginia v. Black*, 538 U.S. 343 (2003).

93. *Gertz v. Robert Welch, Inc.*, 418 U.S. 323 (1974).

94. David Langue and H. Jefferson Power, *No Law: Intellectual Property in the Image of an Absolute First Amendment* (Stanford, CA: Stanford Law Books, 2008).

95. *Harper & Row v. Nation Enterprises*, 471 U.S. 539 (1985). The First Amendment does prevent, however, the use of a prior restraint—an injunction before adjudication about the intellectual property claim, leaving damages and post-publication injunctions as the dominant sanction for violations. Mark Lemley and Eugene Volokh, "Freedom of Speech and Injunctions in Intellectual Property Cases," *Duke Law Journal* 48 (1998).

96. Yochai Benkler, "Free as the Air to Common Use: First Amendment Constraints on Enclosure of the Public Domain," *New York University Law Review* 74 (1999): 354; Neil Weinstock Netanel, "Locating Copyright Within the First Amendment Skein," *Stanford Law Review* 54 (2001): 1; Rebecca Tushnet, "Copyright as a Model for Free Speech Law: What Copyright Has in Common with Anti-Pornography Laws, Campaign Finance Reform, and Telecommunications Regulation," *Boston College Law Review* 42 (2000): 60–63; Eugene Volokh, "Freedom of Speech and Intellectual Property: Some Thoughts After *Elred*, *44 Liquormart*, and *Bartnicki*," *Houston Law Review* 40 (2003): 690.

97. Office of the General Counsel, Harvard University, "Copyright and Fair Use," https://ogc.harvard.edu/pages/copyright-and-fair-use; Baylor University, "Using Copyrighted Material," https://www.baylor.edu/copyright/index.php?id=56543.

98. American Bar Association Business Law Section, "Posting Third-Party Content," https://www.americanbar.org/groups/business_

law/migrated/safeselling/content; "The Digital Millennium Copyright Act: Scope, Reach, and Safe Harbors," *National Law Review*, May 15, 2019, https://www.natlawreview.com/article/digital-millennium-copyright-act-scope-reach-and-safe-harbors.

99. See Sherman Antitrust Act of 1890, 15 U.S.C. §§ 1–7; the Federal Trade Commission Act, 14 U.S.C. § 45; and the Clayton Antitrust Act, 15 U.S.C. §§ 122729 U.S.C. §§ 52–53.

100. See Victor Pickard, *America's Battle for Media Democracy: The Triumph of Corporate Libertarianism and the Future of Media Reform* (New York: Cambridge University Press, 2015), chap. 2.

101. *Associated Press v. United States*, 326 U.S. 1 (1945).

102. *Miami Herald Publishing Co. v. Knight Newspapers*, 418 U.S. 241, 259, 251 n. 17 (1974) (citing Archibald Macleish) (italics omitted).

103. *Red Lion Broadcasting Co.*, at 393 (citing *Associated Press*).

104. *Associated Press v. United States*, 326 U.S. 1 (1945); *Mayee v. Plains Publishing Co.*, 327 U.S. 178 (1946); *FCC v. National Citizens Committee for Broadcasting*, 436 U.S. 755 (1978), *Red Lion Broadcasting Co.*, 395 U.S. (1969).

105. Baker, "*Turner Broadcasting*: Content-Based Regulation of Persons or Presses," 108.

106. Alexander Meiklejohn, *Free Speech and Its Relation to Self-Government* (New York: Harper, 1948), 25.

107. *Miami Herald Publishing v. Knight Newspapers*, 418 U.S. 241, 251 (1974) (citing *Associated Press v. United States*, 326 U.S. 1 (1945).

108. *Associated Press v. United States*, 326 U.S. 1, 20 (1945).

109. Mark Epstein, "Antitrust, Free Speech and Google," *Wall Street Journal*, June 9, 2019, https://www.wsj.com/articles/antitrust-free-speech-and-google-11560108712-.

110. *Miami Herald v. Tornillo*, 418 U.S. 241 (1974).

111. *Turner Broadcasting System v. FCC*, 520 U.S. 180 (1997).

112. See the discussion of the "state action" doctrine later in this chapter. And see Cass Sunstein, *Republic.com* (Princeton, NJ: Princeton University Press, 2001); Cass Sunstein, *Republic.com 2.0* (Princeton, NJ: Princeton University Press, 2009).

113. See David Lyons, "The First Amendment Red Herring in the Net Neutrality Debate," *Forbes*, Mar. 10, 2017, https://www.forbes.com/sites/washingtonbytes/2017/03/10/the-first-amendment-red-herring-in-the-net-neutrality-debate/#65ab7548326a; Dan Kennedy, "The End

of Net Neutrality Will Cripple the First Amendment," WGBH News Commentary, Nov. 27, 2017, https://www.wgbh.org/news/2017/11/27/end-net-neutrality-will-cripple-first-amendment. Contrary views include Brett A. Shumate and Schuyler J. Shouten, "Net Neutrality Proposals for Tech Platforms Raise First Amendment Concern," Jones Day, Sept. 2019, https://www.jonesday.com/en/insights/2019/09/net-neutrality-proposals-for-tech-platforms.

114. Russell Newman, *The Paradoxes of Network Neutralities* (Cambridge, MA: MIT Press, 2019).

115. FCC, "Open Internet Order," 2015; *United States Telecom Association v. Federal Communications Commission and United States of America*, *U.S. Telecom Assn. v. Fed. Commun.*, 855 F.3d 381 (D.C. Cir. 2017) (denying petition for rehearing); *U.S. Telecom Ass'n v. FCC*, 139 S. Ct. 475 (2018) (declining Supreme Court review).

116. Keith Collins, "Net Neutrality Has Officially Been Repealed," *New York Times*, June 11, 2018, https://www.nytimes.com/2018/06/11/technology/net-neutrality-repeal.html.

117. Victor Pickard and David Elliot Berman, *After Net Neutrality: A New Deal for the Digital Age* (New Haven, CT: Yale University Press, 2019). On the upcoming prospects, see Darrell M. West, Nicol Turner Lee, and Caitlin Chin, "What to Expect from a Biden FCC on Section 230, net neutrality, and 5G," Brookings Institution, Dec. 3, 2020, https://www.brookings.edu/blog/techtank/2020/12/03/what-to-expect-from-a-biden-fcc-on-section-230-net-neutrality-and-5g/.

118. The D.C. Circuit denied a petition for rehearing en banc, and the Supreme Court declined to review the court of appeals' June 2016 decision in *United States Telecom Association v. Federal Communications Commission and United States of America*, 855 F.3d 381 (D.C. Cir. 2017), which upheld the net neutrality rule adopted by the FCC. See *United States Telecom Assoc. v. FCC*, No. 15-1063 (D.C. Cir. 2016).

119. *United States Telecom Assoc. v. FCC*, No. 15-1063 (D.C. Cir. 2016), at 528 (Kavanaugh, J., dissenting).

120. *United States Telecom Assoc. v. FCC*, No. 15-1063 (D.C. Cir. 2016), at 388 (Srinivasan, J., concurring).

121. *United States Telecom Assoc. v. FCC*, No. 15-1063 (D.C. Cir. 2016), at 391.

122. Genevieve Lakier, "The Limits of Antimonopoly Law as a Solution to the Problems of the Platform Public Sphere," Knight First Amendment Institute, March 30, 2020, https://knightcolumbia.org/content/the-limits-of-antimonopoly-as-a-solution-to-the-problems-of-the-platform-public-sphere.

123. Brian Stetler, "AT&T CEO Pledges Journalistic Independence for CNN," *Money*, Oct. 23, 2016, https://money.cnn.com/2016/10/23/media/cnn-independence-att-ceo-time-warner.

124. See Ernesto Falcon, "Victory! California's Legislature Pulls AT&T and Comcast Bill That Protected Their Monopolies," Electronic Frontier Foundation, Sept. 10, 2019, https://www.eff.org/deeplinks/2019/09/victory-californias-legislature-pulls-att-and-comcast-bill-protected-their; State of Georgia Public Service Commission, "Who Provides Telecommunication Services?," https://psc.ga.gov/about-the-psc/#telecom.

125. See Susan Crawford, *Captive Audience: The Telecom Industry and Monopoly in the New Gilded Age* (New Haven, CT: Yale University Press, 2013).

126. *Pacific Gas & Electric v. Public Utilities Commission*, 475 U.S. 1 (1986).

127. See *Packingham v. North Carolina*, 582 U.S. ___ (2017), 137 S. Ct. 1730 (2017).

128. See Gregory S. Asciolla, "Note: *Leathers v. Medlock*: Differential Taxation of the Press Survives Under the First Amendment," *Catholic University Law Review* 41 (1992): 507, https://scholarship.law.edu/lawreview/vol41/iss2/6.

129. *Leathers v. Medlock*, 499 U.S. 439 (1991).

130. *Grosjean v. American Press Co.*, 297 U.S. 233 (1936).

131. *Arkansas Writers' Project, Inc. v. Ragland*, 481 U.S. 221 (1987).

132. *Grosjean v. American Press Co.*, 297 U.S. 233 (1936).

133. *Leathers v. Medlock*, 499 U.S. 439 (1991).

134. *Speiser v. Randall*, 357 U.S. 513 (1958).

135. *National Endowment for the Arts v. Finley*, 524 U.S. 569 (1998).

136. 539 U.S. 194 (2003).

137. Nancy Kranich, "Why Filters Won't Protect Children or Adults," *Library Administration and Management* 18 (Winter 2004): 14, ww.ala.org/advocacy/intfreedom/filtering/whyfilterswontprotect; Mary Minow and

Tomas A. Lipinski, *The Library's Legal Answer Book* (Chicago: American Library Association, 2004), 305–306.

138. Compare *Rust v. Sullivan*, 500 U.S. 173 (1991) (allowing restriction on communications about abortion) with *Legal Services Corporation v. Velazquez*, 531 U.S. 533 (2001) (legal services).

139. David Cole, "Beyond Unconstitutional Conditions: Charting Spheres of Neutrality in Government-Funded Speech," *New York University Law Review* 67 (1992): 682; Robert Post, "Subsidized Speech," *Yale Law Journal* 106 (1996): 151; Kathleen M. Sullivan, "Unconstitutional Conditions," *Harvard Law Review* 102 (1989): 1415.

140. *Federal Communications Commission v. League of Women Voters*, 468 U.S. 364 (1984).

141. Martha Minow, "Alternatives to the State Action Doctrine in the Era of Privatization, Mandatory Arbitration, and the Internet: Directing Law to Serve Human Needs," *Harvard Civil Rights–Civil Liberties Law Review* 52 (2017): 146.

142. See the *Civil Rights Cases*, 109 U.S. 3, 25–26 (1883); *PruneYard Shopping Center v. Robbins*, 447 U.S. 74 (1980); *Mazdabrook Commons Homeowners Association v. Khan*, 210 N.J. 482, 493 (2012).

143. Michael J. Phillips, "The Inevitable Incoherence of Modern State Action Doctrine," *St. Louis Law Journal* 28 (1984): 718–721; "Developments in the Law—State Action and the Public/Private Distinction," *Harvard Law Review* 1248 (2010): 1258–1266.

144. *Burton v. Wilmington Parking Authority*, 365 U.S. 715 (1961) (entanglement); *Marsh v. Alabama*, 326 U.S. 501 (1946) (public function). Insufficient to establish state action are instances of private parties subsidized by government, *Jackson v. Metropolitan Edison Co.*, 419 U.S. 345 (1974) or has a government license, see *Rendell-Baker v. Kohn*, 457 U.S. 830 (1982); *Moose Lodge No. 107 v. Irvis*, 407 U.S. 163 (1972).

145. Lillian BeVier and John Harrison, "The State Action Principle and Its Critics," *Virginia Law Review* 96 (2010): 1811, 1827, 1835; Mark Tushnet, 'The Issue of State Action / Horizontal Effect in Comparative Constitutional Law," *International Journal of Constitutional Law* 1 (2003): 79.

146. *Prager University v. Google LLC*, 2018 WL 1471939 (N.D. Cal., March 26, 2018), aff'd (Feb. 26, 2020), 951 F.3d 991 (2020), https://cdn.ca9.uscourts.gov/datastore/opinions/2020/02/26/18-15712.pdf.

147. *Prager University v. Google LLC*, 951 F.3d at 995.

148. *Johnson v. Twitter, Inc.*, No 18CECG00078 (Cal. Super. Ct. June 6, 2018), 9 (permitting ban of high-level white supremacists from the platform). Brian Fung, "Twitter Bans President Trump Permanently," CNN, Jan. 8, 2021, https://www.cnn.com/2021/01/08/tech/trump-twitter-ban/index.html. Such decisions expose chronic questions about the scope and application of company content moderation policies and unpopular decisions may drive users to rivals. Tae Kim, "Can Twitter Survive Banning Donald Trump?" Bloomberg, Jan. 12, 2021, https://www.bloomberg.com/opinion/articles/2021-01-11/can-twitter-survive-banning-donald-trump? Pressure on the companies to make their decisions transparent respects their free speech rights as well as public needs for accountability. Tom Wheeler, "The Consequences of Social Media's Giant Experiment," Brookings Institution, Jan. 13, 2021, https://www.brookings.edu/blog/techtank/2021/01/13/the-consequences-of-social-medias-giant-experiment/.

149. Donie O'Sullivan, "Facebook Takes Down Trump's 'Antifa' Ads 'For Violating Our Policy Against Organized Hate,'" *Mercury News*, June 18, 2020, https://www.mercurynews.com/2020/06/18/facebook-takes-down-trump-ads-for-violating-our-policy-against-organized-hate; Jeff Horwitz and Deepa Seetharaman, "Facebook Executives Shut Down Efforts to Make the Site Less Divisive," *Wall Street Journal*, May 26, 2020; Lila MacLellan, "There Will Never Be Another Facebook Post from Donald Trump as US President, Quqrtz at Work," Jan. 7, 2021, https://qz.com/work/1954112/us-president-donald-trump-will-never-post-on-facebook-again/. Facebook has sent objections over its decision to close off Trump's access to the Facebook Oversight Board.

150. *Adventures Worldwide, LLC v. Google, Inc.*, No. 2:14-cv-646-FtM-PAM-CM, 2017 WL 2210029 (M.D. Fla. Feb. 8, 2017) (holding that Google's delisting a competitor is protected by the First Amendment); *Zhang v. Baidu Com.*, 10 F. Supp. 3d 433, 436-77 (S.D.N.Y. 2014) (results in a search engine are protected by the First Amendment).

151. Twitter (TWTR) Inc., SEC Filing, 10-K Annual Report for the Fiscal Year Ending Tuesday, Dec. 31, 2019, https://last10k.com/sec-filings/twtr#1_0_46; https://business.twitter.com/end-solutions/twitter-ads/amplify-pre-roll.html.

152. *Dreamstime.com, LLC v. Google, LLC*, No. C 18-01910 WHA (N.D. Cal. Jan. 28, 2019), https://casetext.com/case/dreamstimecom-llc-v-google-llc. Dreamstime unsuccessfully alleged an antitrust violation, pointing to

the contract between one of its competitors and Google and claiming the two conspired to injure Dreamstime.

153. *In re Yahoo! Inc. Customer Data Sec. Breach Litig.*, 313 F. Supp. 3d 1113, 1132 (N.D. Cal. 2018).

154. See Jack M. Balkin, "Free Speech in the Algorithmic Society: Big Data, Private Governance, and New School Speech Regulation," *University of California Davis Law Review* 51 (2018). Facebook explains some of its own rules: "About Ads About Social Issues, Elections or Politics," Facebook, May 22, 2020, https://www.facebook.com/business/help/167836590566506?id=288762101909005.

155. See *Song Fi, Inc. v. Google, Inc.*, No. 14-CV-05080-CW, 2016 WL 1298999, at *7 (N.D. Cal. Apr. 4, 2016) (concerning a fraud claim where plaintiffs alleged that defendants had a duty to disclose based on the terms of service contract between parties).

156. *Cyber Promotions v. Am. Online*, 948 F. Supp. 436, 447 (E.D. Pa. 1996).

157. *See* Lincoln Caplan, "Should Facebook and Twitter Be Regulated Under the First Amendment?," Wired, Oct. 11, 2017, https://www.wired.com/story/should-facebook-and-twitter-be-regulated-under-the-first-amendment (quoting Larry Kramer).

158. Id.; see generally Zeynep Tufekci, "It's the (Democracy-Poisoning) Golden Age of Free Speech," *Wired*, Jan. 16, 2018, https://www.wired.com/story/free-speech-issue-tech-turmoil-new-censorship.

159. Jerome Barron, "Access to the Press—a New First Amendment Right," *Harvard Law Review* 80 (1967): 1641; Thomas I. Emerson, "The Affirmative Side of the First Amendment," *Georgia Law Review* 15 (1981): 795.

160. *Associated Press v. United States*, 326 U.S. 1, 20–21 (1945).

161. See "Bush: Bailout Necessary to Deal with Crisis," CNN, Sept. 25, 2008, http://www.cnn.com/2008/POLITICS/09/24/bush.bailout/index.html; see also Josh Zumbrun, "Financial Crisis, Regulatory Agenda Shaped Obama's Economic Legacy," *Wall Street Journal*, Jan. 18, 2017, https://www.wsj.com/articles/financial-crisis-regulatory-agenda-shaped-obamas-economic-legacy-1484762499.

162. Michael Waldman, "The First Step to Hack-Proofing Our Elections," Brennan Center, Feb. 15, 2018, https://www.brennancenter.org/analysis/first-step-hack-proofing-our-elections.

163. First Amendment constraints apply where the government opens up space for expressive purposes, and the government can

limit the forum to certain classes or types of speech. *Good News Club v. Milford Central School*, 533 U.S. 98 (2001). Two courts have found such is the case when President Trump used his Twitter account for communicating broadly, and hence he could not block particular users. *Knight First Amendment Inst. at Columbia Univ. v. Trump*, No. 1:17-cv-5205 (S.D.N.Y.), No. 18-1691 (2d Cir.), No. 20-197 (Supreme Court). See Harold Brubaker, "Elected Officials, Social Media, and the First Amendment," *Philadelphia Inquirer*, Sept. 7, 2017, http://www.philly.com/philly/business/law/elected-officials-social-media-and-the-firstamendment-20170907.html; Anne E. Marmiow, "Trump Cannot Block Critics on Twitter," *Washington Post*, March 23, 2020, https://www.washingtonpost.com/local/legal-issues/trump-cannot-block-critics-on-twitter-federal-court-affirms-in-ruling/2020/03/23/83ac302c-6d0b-11ea-a3ec-70d7479d83f0_story.html.

164. *Packingham v. North Carolina*, 137 S. Ct. 1730, 1735 (2017).

165. 326 U.S. 501 (1946).

166. See *Manhattan Community Access Corp. v. Halleck*, 587 U.S. ___ (2019) (reaffirming state action doctrine in the context of private nonprofit operator of cable channels).

167. See *Johnson v. Twitter, Inc.*, No. 18CECG00078 (Ca. Super. Ct. June 6, 2018); Kate O'Flaherty, "Facebook COVID-19 Fallout: Why Is the Social Network Taking Down Legitimate Posts?," *Forbes*, March 18, 2010, https://www.forbes.com/sites/kateoflahertyuk/2020/03/18/covid-19-fallout-why-is-facebook-wrongly-removing-legitimate-content/#478ed3ba78b4.

168. See Evelyn Douek, "European Commission Communication on Disinformation Eschews Regulation. For Now," Lawfare, May 2, 2018, https://www.lawfareblog.com/european-commission-communication-disinformation-eschews-regulation-now; "Eroding Exceptionalism: Internet Firms' Legal Immunity Is Under Threat," *The Economist*, Feb. 11, 2017, https://www.economist.com/business/2017/02/11/internet-firms-legal-immunity-is-under-threat.

169. https://www.theatlantic.com/technology/archive/2014/06/everything-we-know-about-facebooks-secret-mood-manipulation-experiment/373648/.

170. Linda A. Klein, "A Free Press Is Necessary for a Strong Democracy," *ABA Journal*, May 1, 2017, https://www.abajournal.com/magazine/article/free_press_linda_klein.

171. See Marc Morjé Howard, *The Weakness of Civil Society in Post-Communist Europe* (Cambridge: Cambridge University Press, 2003).

172. "Local Newspapers Are Dying: If Local News Goes Out of Business, the Damage to Our Democracy Will Be Severe and Irreversible," editorial, *Los Angeles Times*, May 24, 2020, https://www.latimes.com/opinion/story/2020-05-24/local-newspapers-dying-ways-to-save-them.

173. See generally Gillian Metzger, "The Constitutional Duty to Supervise," *Yale Law Journal* 124 (2015): 1836; Gillian E. Metzger, "The Supreme Court 2016 Term—Foreword: 1930s Redux: The Administrative State Under Siege," *Harvard Law Review* 131 (2017): 1; David P. Currie, "Positive and Negative Constitutional Rights," *University of Chicago Law Review* 53 (1986): 864; Hope M. Babcock, "The Federal Government Has an Implied Moral Constitutional Duty to Protect Individuals from Harm Due to Climate Change: Throwing Spaghetti Against the Wall to See What Sticks," Georgetown Law School, 2019, https://scholarship.law.georgetown.edu/facpub/2183; Susan Bandes, "The Negative Constitution: A Critique," *Michigan Law Review* 88 (1990): 2271, https://cyber.harvard.edu/vaw00/bandes.html.

174. As thoroughly explored in a recent book, the Commission on Freedom of the Press, led by Robert Maynard Hutchins, in its 1947 report, "A Free and Responsible Press," maintained at times that the First Amendment means, "Congress shall make law enhancing the freedom of speech." Stephen Bates, *An Aristocracy of Critics: Luce, Hutchins, Niebuhr, and the Committee That Redefined Freedom of the Press* (New Haven, CT: Yale University Press, 2020), 5, 174–175, 102, 106. Scholars William Ernest Hocking, Alexander Meiklejohn, and Jerome Barron developed thorough arguments that an affirmative First Amendment obliges government action to reduce monopolies and promote competition, impose duties of common carriage, and subsidize journalism. More recently, the Knight Commission on Trust, Media and Democracy proposed in its 2019 report new duties on journalists as well as more funding by philanthropy, and possibly by government. Knight Foundation, "Crisis in Democracy: Renewing Trust in America," 2019, https://knightfoundation.org/reports/crisis-in-democracy-renewing-trust-in-america.

175. K. Sabeel Rahman, *Democracy Against Domination* (New York: Oxford University Press, 2017), 114.

176. See Ganesh Sitaraman and Anne L. Alstott, *The Public Option: How to Expand Freedom, Increase Opportunity, and Promote Equality* (Cambridge, MA: Harvard University Press, 2019), 224.

CHAPTER FOUR

1. C. West Churchman, "Wicked Problems," *Management Science* 14 (1967): B141–146.

2. See Richard D. Brown, *The Strength of a People: The Idea of An Informed Citizenry in America, 1650–1879* (Chapel Hill: University of North Carolina Press, 1996); Robert W. McChesney and John Nichols, *The Death and Life of American Journalism: The Media Revolution That Will Begin the World Again* (Philadelphia: Nation Books, 2010), 156–212.

3. See Evelyn Douek, "What Kind of an Oversight Board Have You Given Us?," University of Chicago Law Review Blog, May 22, 2020, https://lawreviewblog.uchicago.edu/2020/05/11/fb-oversight-board-edouek Sheera Frankel, "Facebook Bans Holocaust Denial Content," *New York Times*, Oct. 12, 2020, https://www.nytimes.com/2020/10/12/technology/facebook-bans-holocaust-denial-content.html; Salvador Rodriguez, "Facebook Will Ban Ads That Seek to Delegitimize US Election," CNBC, Sept. 20, 2020, https://www.cnbc.com/2020/09/30/facebook-will-ban-ads-that-seek-to-delegitimize-us-election-.html. A real test of who legitimately governs social media concerns Facebook and Donald Trump. See "Trump Wants Back on Facebook. This Star-Studded Jury Might Let Him," *Wired*, Sept. 5, 2020, https://www.wired.com/story/facebook-and-the-folly-of-self-regulation.

4. "'Who Shared It?': How Americans Decide What News to Trust on Social Media," American Press Institute, Mar. 20, 2017, https://www.americanpressinstitute.org/publications/reports/survey-research/trust-social-media.

5. See generally Lee Hollaar, "Legal Protection of Digital Information," Digital Law Online, http://digital-law-online.info/lpdi1.0/treatise33.html, last visited Oct. 25, 2018; Digital Millennium Copyright Act, Pub. L. No. 105–304, 112 Stat. 2860 (codified as amended in scattered sections of 17 U.S.C. and 28 U.S.C.).

6. See Jonathan Taplin, "Is It Time to Break Up Google?," *New York Times*, Apr. 22, 2017, https://www.nytimes.com/2017/04/22/opinion/sunday/is-it-time-to-bbbreak-up-google.html. Users, too, could be required to pay for content. Pryan Chittum, "Mo Pageviews, Mo Problems," Traffic, Sept. 22, 2017, https://traffic.piano.io/2016/09/22/mo-pageviews-moproblems; David Beard, "Spotify for News? Subscription Service Scroll Has New Investor, Partners," Poynter, Feb. 22, 2018, https://www.poynter.org/news/spotify-news-subscription-service-scroll-has-newinvestor-partners. More challenging would be rules that simultaneously require payment and must-carry obligations, for example, regarding local news. See *Turner Broadcasting v. Federal Communications Commission (II)*, 520 U.S. 180 (1997) (permitting must-carry rules for cable).

7. Emily Bell, "Tech Platforms Have a Trust Problem. Do They Care?," *Columbia Journalism Review*, June 23, 2020, https://www.cjr.org/tow_center/tech-platforms-have-a-trust-problem-do-they-care.php.

8. Frank Luby, "Top Pricing Consultant Frank Luby Shares Three Rules for Building a Thriving Media Business in the Age of Free Content," Traffic, https://traffic.piano.io/2016/09/22/the-price-is-wrong, last visited Sept. 21, 2018.

9. See generally Paul Boutin, "The Age of Music Piracy Is Finally Over," Wired, Nov. 29, 2010, https://www.wired.com/2010/11/st-essay-nofreebird; Paul Fingas, "Spotify Really Does Reduce Music Piracy, but at a Cost," Engadget, Oct. 28, 2015, https://www.engadget.com/2015/10/28/spotify-piracy-study; Mark Wilson, "Apple's iTunes Match Legitimizes Music Piracy—Because Piracy No Longer Matters," *Popular Mechanics*, Nov. 15, 2011, https://www.popularmechanics.com/technology/gadgets/a7292/apples-itunes-match-legitimizes-music-piracy-because-piracy-nolonger-matters-6562084. See "Music Piracy Increasing Globally," *South China Post*, Sept. 20, 2017, http://www.scmp.com/culture/music/article/2112017/music-piracyincreasing-globally-ripped-spotify-youtube-says-recording.

10. See Benjamin Mullin, "Google Pledges $1 Billion in Licensing Payments to News Publishers," *Wall Street Journal*, Oct. 1, 2020; Gerry Smith, "Publishers See 2020 as the Year More Start to Get Paid for News," Bloomberg, Dec. 23, 2019, https://www.bloomberg.com/news/articles/2019-12-23/publishers-see-2020-as-the-year-more-start-to-get-paid-for-news. Nonetheless, Google threatened to withdraw from Australia as the country planned to enforce its requirement of paying publishers for links to news

searches; Mike Cherney, "Google Threatens to Cut Off Australia," *Wall Street Journal*, Jan. 22, 2021, https://www.wsj.com/articles/google-escalates-dispute-with-australia-by-threatening-search-shutdown-11611298523.

11. Facebook Journalism, "3 Facebook Tools That Connect People with Local Breaking News," Facebook, Sept. 20, 2019, https://www.facebook.com/journalismproject/facebook-tools-local-breaking-news; Jacob Kastrenakes, "Twitter Will Livestream Local Broadcasts During Breaking News Events," The Verge, Feb. 15, 2018, https://www.theverge.com/2018/2/15/17016434/twitter-local-news-broadcasts-breaking-events.

12. Kim Lyons, "Australia Will Compel Facebook and Google to Pay Media Outlets for News Content," The Verge, Apr. 19, 2020, https://www.theverge.com/2020/4/19/21227263/australia-facebook-google-ad-revenue-media-coronavirus-economy; Ben Smith, "Big Tech Has Crushed the News Business: That's About to Change," *New York Times*, May 10, 2020, https://www.nytimes.com/2020/05/10/business/media/big-tech-has-crushed-the-news-business-thats-about-to-change.html?smid=nytcore-ios-share.

13. Julia Fioretti, "EU States Agree Rules to Make Search Engines Pay for News," May 25, 2018, https://www.reuters.com/article/us-eu-copyright/eu-states-agree-rules-to-make-search-engines-pay-for-news-idUSKCN1IQ2NS.

14. Directive of the European Parliament and of the Council of the European Union on Copyright and Related Rights in the Digital Single Market, Apr. 17, 2019, https://eur-lex.europa.eu/legal-content/EN/TXT/PDF/?uri=CELEX:32019L0790&from=EN. Working out compliance with copyright laws, freedom of expression, and privacy is a complex challenge, especially with the use of digital content moderation tools, prompting international discussion and court cases. See Julia Reda, Joschka Selinger, and Michael Servatius, "Article 17 of the Directive on Copyright in the Digital Single Market: A Fundamental Rights Assessment," Gesellschaft für Freiheitsrechte e.V. / Society for Civil Rights, Berlin, Nov. 16, 2020.

15. "This Day in History, ASCAP Is Founded," History Channel, https://www.history.com/this-day-in-history/ascap-is-founded.

16. See Andrew P. Bridges, Crystal Nwaneri, Jennifer Stanley, and Chieh Tung, "The EU Copyright Directive: Potential Copyright Liability and 'Best Efforts' Standard for Platforms," Fenwick and West LLP, Jan. 27, 2020, https://www.jdsupra.com/legalnews/the-eu=-copyright-directive-potential-84751?.

17. Johannes Munter, "Google News Shutdown in Spain Was Not as Bad as Google Would Have You Believe," News Media Alliance, Nov. 14, 2019, https://www.newsmediaalliance.org/google-news-shutdown-in-spain-not-as-bad-as-google-would-have-you-believe. See Lucinda Southern, "State of Play: Where the Battle with Google and Facebook to Pay for News is Hottest," *Digiday*, Sept. 7. 2020, digiday.com/media/state-of-play-where-the-battle-with-google-and-facebook-to-pay-for-news-is-hottest/.

18. 47 U.S.C. § 230 (2018) ("No provider or user of an interactive computer service shall be treated as the publisher or speaker of any information provided by another information content provider").

19. 47 U.S.C. § 230 (c)(1)(2018).

20. Danielle Keats Citron and Benjamin Wittes, "The Internet Will Not Break: Denying Bad Samaritans Section 230 Immunity," *Fordham Law Review* 86 (2017): 406–410.

21. *Backpage.com LLC v. McKenna*, 881 F.Supp.2d 1262 (W.D. Wash. 2012); Elizabeth Dias, "Trump Signs Bill Amid Momentum to Crack Down on Trafficking," *New York Times*, Apr. 11, 2018, https://wwwnytimes.com/2018/04/11/us/backpage-sex-trafficking.html?moduleDetail=section-news-3&action-click&contentCollection=Politics&region=Footer&module; Sarah N. Lynch and Lisa Lambert, "Sex Ads Website Backpage Shut Down by U.S. Authorities," Reuters, Apr. 6, 2018, https://www.reuters.com/article/us-usa-backpage-justice/sex-ads-website-backpage-shut-down-by-u-s-authorities-idUSKCN1HD2QP.

22. *Batzel v. Smith*, 333 F.3rd 1018 (9th Cir. 2003); *Carafano v. Metrosplash.com*, 339 F.3rd 1119 (9th Circ. 2003); *Zeran v. AOL*, 129 F.3rd 327 (4th Cir. 1997).

23. *Force v. Facebook, Inc.*, 934 F.3rd 54 (2d Cir. 2019).

24. *Ben Ezra, Weinstein and Co. v. America Online*, 206 F.3d. 90 (10th Cir. 2000), cert. denied, 531 U.S. 824 (2000).

25. *Chicago Lawyers' Committee for Civil Rights Under Law, Inc. v. Craigslist, Inc.*, 519 F.3d 666 (7th Cir. 2008).

26. Mary Anne Franks, "Our Collective Responsibility for Mass Shootings," *New York Times*, Oct. 11, 2019, https://www.nytimes.com/2019/10/09/opinion/mass-shooting-responsibility.html.

27. See Citron and Wittes, "The Internet Will Not Break"; Heather Whitney, "Search Engines, Social Media, and the Editorial Analogy,"

Knight First Amendment Institute, Feb. 2018, https://knightcolumbia.org/content/search-engines-social-media-and-editorial-analogy.

28. John D. McKinnon and Rebecca Ballhaus, "Trump Signs Executive Order Targeting Social Media," *Wall Street Journal*, May 28, 2020, https://www.wsj.com/articles/trump-to-sign-executive-order-targeting-social-media-11590681930?mod=djemalertNEWS. For articulation of its legal defects, see Laurence H. Tribe and Joshua A. Geltzer, "Trump Is Doubly Wrong About Twitter," *Washington Post*, May 28, 2020, https://www.washingtonpost.com/opinions/2020/05/28/trump-is-doubly-wrong-about-twitter.

29. See U.S. Department of Justice, "Justice Department Issues Recommendations for Section 230 Reform," June 17, 2020, https://www.justice.gov/opa/pr/justice-department-issues-recommendations-section-230-reform; David McCabe, "Tech Companies Shift Their Posture on a Legal Shield, Warry of Being Left Behind," *New York Times*, Dec. 15, 2020, https://www.nytimes.com/2020/12/15/technology/tech-section-230-congress.html.

30. Tarleton Gillespie, *Custodians of the Internet* (New Haven, CT: Yale University Press, 2018), 1–44. Even sites claiming to do little moderation do some, and they also rely on other companies to host or distribute, which can refuse to carry based on content, as Parler discovered after it was associated with the January 6, 2021 assault on the U.S. Capitol. Elizabeth Culliford and Jonathan Stempel, "Parler Loses Bid to Require Amazon to Restore Service," *Reuters*, Jan. 21, 2021, reuters.com/article/us-amazon-com-parler/parler-loses-bid-to-require-amazon-to-restore-service-idUSKBN29Q2T3; Jeff Horwitz and Keach Hagey, "Parler Makes Play for Conservatives Mad at Facebook, Twitter," *Wall Street Journal*, Nov. 14, 2020, https://www.wsj.com/articles/parler-backed-by-mercer-family-makes-play-for-conservatives-mad-at-facebook-twitter-11605382430.

31. Robert Gorwa, Reuben Binns, and Christian Katzenbach, "Algorithmic Content Moderation: Technical and Political Challenges in the Automation of Platform Governance," *Big Data and Society* 7 (February 28, 2020), https://journals.sagepub.com/doi/full/10.1177/2053951719897945; Billy Perrigo, "Facebook Says It's Removing More Hate Speech Than Ever Before. But There's a Catch," *Time*, Nov. 27, 2019, https://time.com/5739688/facebook-hate-speech-languages.

32. Danielle Keats Citron, "Attachment—Additional Questions for the Record," presented to the Subcommittee on Communications and

Technology and Subcommittee on Consumer Protection and Commerce Joint Hearing on "Fostering a Healthier Internet to Protect Consumers," October 16, 2019. See also Miguel Casillas, "Section 230 Reform Is Not Dark and Full of Terrors: A Duty of Care to Protect Disadvantaged Groups from Online Hatred," unpublished paper, 2020.

33. Shona Ghosh, "Sheryl Sandberg Just Dodged a Question About Whether Facebook Is a Media Company," Business Insider, Oct. 12, 2017, http://www.businessinsider.com/sheryl-sandberg-dodged-question-on-whether-facebook-is-a-media-company-2017-10; Cecilia Kang and Mike Isaac, "Defiant Zuckerberg Says Facebook Won't Police Political Speech," *New York Times*, Oct. 17, 2019, https://www.nytimes.com/2019/10/17/business/zuckerberg-facebook-free-speech.html.

34. See *Force v. Facebook, Inc.*, cert. denied, *Force v. Facebook, Inc.*, 19-859, 2020 WL 2515485 (U.S. May 18, 2020).

35. Matt Schruers, "Debate over Online Content Embodies 'Moderator's Dilemma,'" DiscO, Sept. 4, 2019, http://www.project-disco.org/innovation/090419-debate-over-online-content-embodies-moderators-dilemma.

36. Chloe Hadavas, "Why We Should Care That Facebook Accidentally Deplatformed Hundreds of Users," *Slate*, June 12, 2020, https://slate.com/technology/2020/06/facebook-anti-racist-skinheads.html.

37. Thanks to Vicki Jackson for this suggestion, based on *Citizens United v. FEC*, 558 U.S. 310, 335–36 (2010).

38. *Jian Zhang v. Baidu.com Inc.*, 10 F.Supp. 3rd 433 (S.D.N.Y. 2014).

39. See Scott Horton, "James Madison and the Corporations," Berfrois, May 6, 2011, https://www.berfrois.com/2011/05/james-madison-and-the-corporations.

40. U.S. Department of Justice, "Attorney General Eric Holder Speaks at the Sherman Act Award Ceremony," Apr. 20, 2010, https://www.justice.gov/opa/speech/attorney-general-eric-holder-speaks-sherman-act-award-ceremony (quoting Senator John Sherman).

41. Richard Hofstadter, "What Happened to the Antitrust Movement?," in *The Business Establishment*, ed. Earl Cheit (New York: Wiley and Sons, 1994), 113; Herbert Hovenkamp, "Antitrust Policy, Federalism, and the Theory of the Firm: An Historical Perspective," *Antitrust Law Journal* 59 (1990): 59.

42. See K. Sabeel Rahman, *Democracy Against Domination* (New York: Oxford University Press, 2017); Tim Wu, *The Curse of Bigness: Antitrust in the New Gilded Age* (New York: Columbia Global

Reports, 2018); Lina Khan, "The New Brandeis Movement: America's Antimonopoly Debate," *Journal of European Competition Law and Practice* 9 (Mar. 2018): 131, https://academic.oup.com/jeclap/article/9/3/131/4915966; Ganesh Sitaraman, "Regulating Tech Platforms: A Blueprint for Reform," Great Democracy Initiative, Vanderbilt Law School, Research Paper No. 18-64, Apr. 2018, https://ssrn.com/abstract=3278418; Dina Srinivasan, "The Antitrust Case Against Facebook," *Berkeley Business Law Journal* 16 (2019): 39; Tim Wu, "Blind Spot: The Attention Economy and the Law," *Antitrust Law Journal*, 2018, https://scholarship.law.columbia.edu/faculty_scholarship/2029. See also Joseph Coniglio, "Why the 'New Brandeis Movement' Gets Antitrust Wrong," Law360, Apr. 24, 2018, https://www.law360.com/articles/1036456/why-the-new-brandeis-movement-gets-antitrust-wrong; Thomas A. Piraino Jr., "Reconciling the Harvard and Chicago Schools: A New Antitrust Approach for the 21st Century," *Indiana Law Journal* 82 (2007): 345, http://www.repository.law.indiana.edu/ilj/vol82/iss2/4.

43. Brian Fung, "Facebook Must Be Broken Up, the US Government Says in Groundbreaking Lawsuit," CNN, Dec. 9, 2020, https:www.cnn.com/12/09/facebook-antitrust-lawsuit-ftc-attorney-generals/index.html?; Cecilia Kang and Mike Isaac, "U.S. and States Say Facebook Illegally Crushed Competition," *New York Times*, Dec. 9, 2020, https://www.nytimes,com/2020/12/09/technology/facebook-antitrust-monopoly.html; David McLaughlin, Ben Brody, and Kurt Wagner, "Facebook Latest FTC Trouble: Social Media, Advertising Probe," Bloomberg Law, July 25, 2019, https://news.bloomberglaw.com/tech-and-telecom-law/facebook-latest-ftc-headache-probe-of-social-media-competition. 6.

44. Siva Vaidhyanathan, *Anti-Social Media: How Facebook Disconnects Us and Undermines Democracy* (New York: Oxford University Press, 2018), 102.

45. See Tim Wu, *The Attention Merchants: The Epic Scramble to Get Inside Our Heads* (New York: Knopf, 2016); Wu, "Blind Spot."

46. Wu, "Blind Spot."

47. Jessica Hullinger, "Why Social Media Deliberately Manipulate Your Emotions," *Fortune*, Dec. 20, 2015, https://fortune.com/2015/12/30/social-media-emotions. But see Addiction Resource, "Social Media Addiction: The Facts and Solutions," https://addictionresource.com/addiction/technology-addiction/social-media-addiction.

48. Annemarie Bridy, "Remediating Social Media: A Layer-Conscious Approach," *Boston University Journal of Science and Technology Law* 24 (2018): 193, 194.

49. Tom Wheeler, *From Gutenberg to Google: The History of Our Future* (Washington, DC: Brookings Institution, 2019), 213–214; Sally Hubbard, "The Case for Why Big Tech Is Violating Antitrust Laws," CNN, Jan. 2, 2019, https://www.cnn.com/2019/01/02/perspectives/big-tech-facebook-.

50. Eyal Benvenisti, "Upholding Democracy Amid the Challenges of New Technology: What Role for the Law of Global Governance?," *European Journal of International Law* 29, no. 1 (2018): 62.

51. Jason Murdock, "Facebook Is Tracking You Online, Even If You Don't Have an Account," *Newsweek*, Apr. 17, 2018, https://www.newsweek.com/facebook-tracking-you-even-if-you-do-not-have-account-888699.

52. Paul Hitlin and Lee Rainie, "Facebook Algorithms and Personal Data," Pew Research Center: Internet and Technology, Jan. 16, 2019, https://www.pewinternet.org/2019/01/16/facebook-algorithms-and-personal-data. See also Srinivasan, "The Antitrust Case Against Facebook."

53. "Taming the Titans," *The Economist*, Jan. 20, 2018.

54. Eric Posner and E. Gen Weyl, *Radical Markets: Uprooting Capitalism and Democracy for a Just Society* (Princeton, NJ: Princeton University Press, 2018; David Scharfenberg, "Facebook Should Pay You for Your Data," *Boston Globe*, June 17, 2018.

55. "Bundeskartellamt Prohibits Facebook from Combining User Data from Different Sources," Bundeskartellamt, Feb. 7, 2019, https://www.bundeskartellamt.de/SharedDocs/Meldung/EN/Pressemitteilungen/2019/07_02_2019_Facebook.html. See Giuseppe Colangelo, "Data Accumulation and the Privacy-Antitrust Interface: Insights from the Facebook Case," *International Data Privacy Law* 8 (Aug. 2018): 224, https://academic.oup.com/idpl/articles/8/3/224/18967.

56. The federal government and forty-six states launched antitrust suits against Facebook in December 2020, following similar efforts against Google in October 2020; such efforts have support from leaders in both political parties. Fung, "Facebook Must Be Broken Up," supra; "A Formidable Alliance Takes on Facebook," *The Economist* (Dec. 12, 2020), https://www.economist.com/business/2020/12/12/a-formidable-alliance-takes-on-facebook.

57. Gregory Ferenstein, "Mark Zuckerberg Talks Breaking Up Facebook, Elections, and External Oversight," *Forbes*, June 26, 2019, https:www.forbes.com/sites/gregoryferenstein/2019/26/zuckerberg.

58. Katy Steinmetz, "Inside Instagram's War on Bullying," *Time*, July 8, 2019, https://time.com/5619999/instagram-mosseri-bullying-artificial-intelligence. If broken into smaller parts, the new companies could in theory cooperate on some activities, but such cooperation would be sharply constrained by antitrust restrictions.

59. "Cooperation or Resistance?: The Role of Tech Companies in Government Surveillance," *Harvard Law Review* 131 (Apr. 10, 2018), 1722, https://harvardlawreview.org/2018/04/cooperation-or-resistance-the-role-of-tech-companies-in-government-surveillance.

60. Ganesh Sitaraman, "The National Security Case for Breaking Up Big Tech," Knight First Amendment Institute, Jan. 20, 2020, https://knightcolumbia.org/content/the-national-security-case-for-breaking-up-big-tech.

61. Wendy Gordon, "Fair Use as Market Failure: A Structural and Economic Analysis of the 'Betamax' Case and Its Predecessors," *Columbia Law Review* 82 (1982): 1600–1657.

62. Veronica Marotta, Vibhanshu Abhishek, and Alessandro Acquisti, "Online Tracking and Publishers' Revenues: An Empirical Analysis," unpublished paper, May 2019, available at https://weis2019.econinfosec.org/wp-content/uploads/sites/6/2019/05/WEIS_2019_paper_38.pdf; Jason Kint, "Behavioral Advertising: The Mirage Built by Google," Digital Content Next, June 6, 2019, https://digitalcontentnext.org/blog/2019/06/06/behavioral-advertising-the-mirage-built-by-google.

63. K. Sabeel Rahman, "The New Utilities: Private Power, Social Infrastructure, and the Revival of the Public Utility Concept," *Cardozo Law Review* 39 (2018): 1632.

64. Martin G. Glaeser, *Public Utilities in American Capitalism* (New York: Macmillan, 1957), 33.

65. Rahman, "The New Utilities," 1635; K. Sabeel Rahman, "Infrastructural Regulation and the New Utilities," *Yale Journal on Regulation* 35 (2018) (quoting Louis Brandcis). Scc also Nicholas Bagley, "Medicine as a Public Calling," *Michigan Law Review* 114 (2015): 75.

66. See *New State Ice Co. v. Liebmann*, 285 U.S. 262, 277–278 (1932) (Brandeis, J., dissenting).

67. Rahman, "Infrastructural Regulation and the New Utilities," 914–916. The internet platform companies operate at such scale and impact that retrospective responses to distinct problems cannot tackle issue the way prospective guidelines would. Drew Margolin, "First Principles," Multichannel News, Sept. 12, 2018, https://www.nexttv.com/blog/first-principles.

68. dana boyd, "Facebook Is a Utility and Utilities Get Regulated," Apophenia, May 15, 2010, http://www.zephoria.org/thoughts/archives/2010/05/15/facebook-is-a-utility-utilities-get-regulated.html. See also Zeynep Tufekci, "Google Buzz: The Corporation of Social Commons," Technosociology, Feb. 17, 2010, http://technosociology.org/?p=102.

69. See "Cooperation or Resistance: The Role of Technology Companies in Government Surveillance," *Harvard Law Review* 131 (2018): 1722; Kif Leswing, "New York City Is Turning On Its Google-Owned Free Wi-Fi Network," *Fortune*, Jan. 5, 2016, https://fortune.com/201601/05/link-nyc-unveiled.

70. Julie E. Cohen, "Law for the Platform Economy," *University of California Davis Law Review* 51 (2017): 143–145. See also José Van Dijck, Thomas Poell, and Martijn de Waal, *The Platform Society: Public Values in a Connective World* (New York: Oxford University Press, 2018), 12–30.

71. Tim Wu, "Network Neutrality, Broadband Discrimination," *Journal of Telecommunications and High Technology Law* 2 (2003): 141. An internet service provider could be tempted to prefer some providers over others, either by willingness to pay or other, more content-oriented reasons. The Obama-era FCC pursued net neutrality; the FCC under President Trump has repealed those efforts. Keith Collins, "Why Net Neutrality Was Repealed and How It Affects You," *New York Times*, Dec. 14, 2017, https://www.nytimes.com/2017/12/14/technology/net-neutrality-rules.html; Brian Fung, "The FCC's Vote Repealing Its Net Neutrality Rules Is Finally Official," *Washington Post*, Feb. 22, 2018, https://www.washingtonpost.com/news/the-switch/wp/2018/02/22/the-fccs-net-neutrality-rules-will-die-on-april-23-heres-what-happens-now/?utm_term=.e05c92ce 7b9f. In January 2018, the governor of Montana signed a bill requiring net neutrality for ISPs doing business in the state, and other states may follow. Cecilia Kang, "Montana Governor Signs Order to Force Net Neutrality," *New York Times*, Jan. 22, 2018, https://www.nytimes.com/2018/01/22/technology/montana-net-neutrality.html. Defenders of net neutrality are also pursuing other efforts. See Tony Romm, "It Ain't Over: Net Neutrality Advocates Are

Preparing a Massive New War Against Trump's FCC," Vox Recode, Jan. 4, 2018, https://www.vox.com/2018/1/4/16846978/net-neutrality-internet-donald-trump-ajit-pai-fcc-democrats-advocates-election.

72. Susan Crawford, "First Amendment Common Sense," *Harvard Law Review* 127 (2015): 2354; Wu, "Network Neutrality, Broadband Discrimination," 169; FCC, "Protecting and Promoting the Open Internet Order," 30 FCC Record 5601, GN Docket No. 14-28 (2015); "Restoring Internet Freedom Order," 33 FCC Record 311, WC Docket No. 17-108 (2018).

73. See K. Sabeel Rahman, "Regulating Informational Infrastructure: Internet Platforms as the New Public Utilities," *Georgetown Law Technology Review* 2 (2018): 234; Rahman, "Infrastructural Regulation and the New Utilities," 911.

74. Firms could also create their own external panels having access to relevant materials and providing fairness audits of firm practices. See Eve Smith, "Silicon Valley, We Have a Problem," *The Economist*, Jan. 20, 2018.

75. Mark Tushnet, "Institutions Protecting Democracy," *University of Toronto Law Review* 70 (Spring 2020): 95.

76. Mathew Ingram and Alex Stamos, "Former Facebook Security Chief Alex Stamos Talks About Political Advertising," Galley by CJR, Jan. 16, 2019, https://galley.cjr.org/public/conversations/-LyjQOoPX4-yK-H78Mw6.

77. "Fake News Expert on How False Stories Spread and Why People Believe Them," *Fresh Air*, NPR, December 14, 2016, https://www.npr.org/2016/12/14/505547295/fake-news-expert-on-how-false-stories-spread-and-why-people-believe-them.

78. Yochai Benkler, Robert Faris, and Hal Roberts, *Network Propaganda Manipulation, Disinformation, and Radicalization* (New York: Oxford University Press, 2018), 371; Amar Bakshi, "Why and How to Regulate Online Advertising in Online News Publications," *Journal of Media Law and Ethics* 4 (2015): 22.

79. Melynda Fuller, "Microsoft to Give Edge Browser Users Access to NewsGuard Tools," Publishers Daily, May 14, 2020, https://www.mediapost.com/publications/article/351424/microsoft-to-give-edge-browser-users-access-to-new.html.

80. Benkler, Faris, and Roberts, *Network Propaganda*, 386. But see Lina Khan and David Pozen, "A Skeptical View of Information Fiduciaries," *Harvard Law Review* 133 (2019): 487.

81. See generally Susan Crawford, *Captive Audience: The Telecom Industry and Monopoly Power in the New Gilded Age* (New Haven, CT: Yale University Press, 2013).

82. Jonathan Zittrain, "How to Exercise the Power You Didn't Ask For," *Harvard Business Review*, Sept. 2018, https://hbr.org/2018/09/how-to-exercise-the-power-you-didnt-ask-for.

83. Thanks to Zachary Meskell for the idea of an "awareness doctrine."

84. "Help You Make Sense of the News," Google News Initiative, June 1, 2020, https://newsinitiative.withgoogle.com/hownewsworks/mission/help-you-make-sense-of-the-news; "Content Policies," Google News Publisher Center Help, June 1, 2020, https://support.google.com/news/publisher-center/answer/6204050?visit_id=637267349198073900-469138961&rd=1.

85. "How Facebook News Works," Facebook, June 2, 2020, https://www.facebook.com/news/howitworks.

86. Lewis Rice, "Common Threat: Sunstein Urges People to Consume More Diverse Information for the Good of Our Democracy," Harvard Law Today, July 25, 2017, https://today.law.harvard.edu/book-review/common-threat; see also Robert Farley, "Cass Sunstein Once Considered a 'Fairness Doctrine' of Sorts for the Internet, but Then Thought Better of It," Politifact, May 5, 2009, http://www.politifact.com/truth-o-meter/statements/2009/may/05/chain-email/cass-sunstein-once-considered-fairnes-doctrine-sor.

87. Frank Pasquale, "Search Neutrality as Disclosure and Auditing," Concurring Opinions, Feb. 19, 2011, https://concurringopinions.com/archives/2011/02/search-neutrality-as-disclosure-and-auditing.html; James Grimmelmann, "Some Skepticism About Search Neutrality," in *The Next Digital Decade: Essays on the Future of the Internet*, ed. Berin Szoka and Adam Marcus (Washington, DC: TechFreedom, 2010).

88. Jamie Ehrlich, "Student Start-Up Series: 'Flipside' Bolsters Political Discourse Through Technology," University of Chicago, Nov. 1, 2017, http://college.uchicago.edu/uniquely-chicago/story/student-start-series-%E2%80%98flipside%E2%80%99-bolsters-political-discourse-through.

89. Telecommunications Act of 1996, 110 Stat. 56 (1996), at section 551(d).

90. There may be lessons from the way individuals receiving information about their food consumption can be nudged toward healthy diets. See Jason Gilliland et al., "Using a Smartphone Application to Promote Healthy

Dietary Behaviours and Local Food Consumption," 2015 BioMed Research Int'l 841368, https://www.ncbi.nlm.nih.gov/pmc/articles/PMC4561980/.

91. See Red Lion Broadcasting Co. v. FCC, 395 U.S. 367 (1969). Court opinioning co. v. c Red Lion Broadcasting Co. v. FCC, 395 U.S. 367 – 196.

92. Senator Josh Hawley, "Senator Hawley Introduces Legislation to Amend Section 230 Immunity for Big Tech Companies," June 19, 2019, https://www.hawley.senate.gov/senator-hawley-introduces-legislation-amend-section-230-immunity-big-tech-companies; "Hawley Proposes a Fairness Doctrine for the Internet," Tech Freedom, June 19, 2019, https://techfreedom.org/hawley-proposes-a-fairness-doctrine-for-the-internet.

93. Mike Masnick, "It's Spreading: Lindsey Graham Now Insisting 'Fairness Doctrine' Applies to the Internet," Tech Dirt, Apr. 22, 2018, https://www.techdirt.com/articles/20180422/17453639689/spreading-lindsey-graham-now-insisting-fairness-doctrine-applies-to-internet.shtml.

94. The same can be said regarding traditional media and ad dollars. See Margaret Sullivan, "Fox News is a Hazard to Our Democracy. It's Time to Take the Fight to the Murdochs. Here's How," *Washington Posti*, Jan. 24, 2021, https://www.washingtonpost.com/lifestyle/media/fox-news-is-a-hazard-to-our-democracy-its-time-to-take-the-fight-to-the-murdochs-heres-how/2021/01/22/1821f186-5cbe-11eb-b8bd-ee36b1cd18bf_story.html?fbclid=IwAR3295y4br7zBaMoGca1mhtDfykiyh_gDkVADsgJIdPPX8lfbCYCgbB_MlI#click=https://t.co/kYnYcbHceU.

95. David Berreby, "Click to Agree with What? No One Reads Terms of Service, Studies Confirm," *The Guardian*, Mar. 3, 2017, https://www.theguardian/com/technology/2017/mar/03/terms-of-service-online-contracts-fine-print.

96. See, e.g., *Fraley v. Facebook, Inc.*, 830 F. Supp. 2d 785, 813-14 (N.D. Cal. 2011)

97. Gabriel Wood, "How Enforceable Are Terms of Service Agreements?," NextAdvisor, Nov. 1, 2017, https://www.nextadvisor.com/how-enforceable-are-terms-of-service-agreements; Lawrence Lessig, "Aaron's Law: Violating a Site's Terms of Service Should Not Land You in Jail," *The Atlantic*, Jan. 16, 2013, https://www.theatlantic.com/technology/archive/2013/01/Aarons-law-violating-a-sites-terms-of-service-should-not-land-you-in-jail/267247.

98. See *Zauderer v. Office of Disciplinary Counsel*, 471 U.S. 626 (1985); *Central Hudson Gas and Electric Corp. v. Public Services Commission*, 447 U.S. 557, 566 (1980).

99. Sam Meredith, "Here's Everything You Need to Know About the Cambridge Analytica Scandal," CNBC, Mar. 21, 2018, https://www.cnbc.com/2018/03/21/facebook-cambridge-analytica-scandal-everything-you-need-to-know.html.

100. See *Sorrell v. IMS*, 564 U.S. 552 (2011).

101. See Stanford Law and Policy Lab Report, "Fake News and Misinformation," Oct. 2017 (student project), https://law.stanford.edu/wp-content/uploads/2017/10/Fake-News-Misinformation-FINAL-PDF.pdf. The report also found that users are more likely to believe content on Twitter that is marked as "verified," although the readers "sometimes conflated the authenticity of accounts associated with people Twitter deems to be 'in the public interest' with accurate information."

102. *McIntyre v. Ohio Elections Commission*, 514 U.S. 334 (1995). Limits on such protection have been allowed in the area of campaign finance: *McConnell v. Federal Election Commission*, 540 U.S. 93 (2003).

103. *State v. Miller*, 260 Ga. 699 (1990).

104. See *United States v. Alvarez*, 567 U.S. 709, 734 (2012) (Breyer, J., concurring in the judgment).

105. Federal Communications Commission, "Broadcasting False Information," https://www.fcc.gov/consumers/guides/broadcasting-false-information.

106. *Parino v. BidRack, Inc.*, 838 F. Supp. 2d. 900, 905–06 (N.D. Cal. 2011).

107. The Supreme Court's restrictions on compelled speech so far have not interfered with mandated disclosures of accurate information for consumers. See *American Meat Institute v. USDA*, 760 F.3rd 18, 22023 (D.C. Cir. 2014) (en banc).

108. See "Fake News Expert on How False Stories Spread and Why People Believe Them."

109. See generally *New York Times Co. v. Sullivan*, 376 U.S. 254 (1964); *United States v. Alvarez*, 567 U.S. 709 (2012).

110. See *People v. Croswell*, 3 Johns. Cas. 337, 393 (N.Y. Sup. Ct. 1804). Actions for civil liability for defamation continue in the mass media era and can proceed against those who propagate conspiracies and misinformation. Colin Kalmbacher, "Texas Supreme Court Silently Denies Alex Jones All Forms of Relief: Sandy Hook Families and Others Can Now Sue Conspiracy Theorist and InfoWars into the Ground," *Law and Crime*,

Jan. 22, 2021, https://lawandcrime.com/high-profile/texas-supreme-court-silently-denies-alex-jones-all-forms-of-relief-sandy-hook-families-and-others-can-now-sue-conspiracy-theorist-and-infowars-into-the-ground/.

111. See generally *Bolger v. Youngs Drug Prods. Corp.*, 463 U.S. 60 (1983); *Cent. Hudson Gas and Elec. Corp. v. Pub. Serv. Comm'n*, 447 U.S. 557 (1980); *Friedman v. Rogers*, 440 U.S. 1 (1979); Gene Quinn, "Does the First Amendment Protect False and Misleading Speech?," IP Watchdog, Feb. 9, 2012, http://www.ipwatchdog.com/2012/02/09/does-the-first-amendment-protect-false-and-misleading-speech/id=22202.

112. Consumers International, "Social Media Scams: Understanding the Consumer Experience to Make a Safer Digital World," May 2019, 20, https://www.consumersinternational.org/media/293343/social-media-scams-final-245.pdf.

113. *Federal Trade Commission v. NPB Advertising, Inc.*, 218 F. Supp. 3rd 1352 (M.D. Fla. 2016); *FTC v. Coulomb Media, Ind.*, 2012 WL 965992, at *4–5 (E.D. Mich. 2012).

114. *Federal Trade Commission v. LeadClick Media*, LLC, 838 F.3rd 158 (ed. Circ. 2016).

115. FCC Media Bureau, *The Public and Broadcasting* 12 (2019), https://www.facc.gov/sites/default/files/public-and-broadcasting.pdf.

116. SMART Act, https://www.congress.gov/bill/116th-congress/senate-bill/2314/text; Casey Newton, "New Legislation Is Putting Social Networks in the Crosshairs," The Verge, Aug. 1, 2019, https://www.theverge.com/2019/8/1/20749517/social-network-legislation-hawley-privacy-research.

117. Yafet Lev-Aretz, "Facebook and the Perils of a Personalized Choice Architecture," TechCrunch, Apr. 24, 2018, https://techcrunch.com/2018/04/24/facebook-and-the-perils-of-a-personalized-choice-architecture.

118. Cass Sunstein, "The Ethics of Nudging," *Yale Journal on Regulation* 32 (2015): 414, 416, 428.

119. Shoshana Zubloff, " 'Surveillance Capitalism' Has Gone Rogue," *Washington Post*, Jan. 24, 2019, https://www.washingtonpost.com/opinions/surveillance-capitalism-has-gone-rogue-we-must-curb-its-excesses/2019/01/24/be463f48-1ffa-11e9-9145-3f74070bbdb9_story.html.

120. Daniel Susser, Beate Roessler, and Helen Nissenbaum, "Online Manipulation: Hidden Influences in a Digital World," *Georgetown Law Technology Review* 4 (2019): 1.

121. Shoshana Zubloff, *The Age of Surveillance Capitalism: The Fight for a Human Future at the Frontier of Power* (New York: PublicAffairs, 2019).

122. See Przemysław Pałka, "The World of Fifty (Interoperable) Facebooks," Mar. 10, 2020, https://papers.ssrn.com/sol3/papers.cfm?abstract_id=3539792.

123. Berreby, "Click to Agree with What?"

124. See Matteo Monti, "Perspectives on the Regulation of Search Engine Algorithms and Social Networks: The Necessity of Protecting the Freedom of Information," *Opinio Juris in Comparatione: Studies in Comparatives and National Law* 1 (2017): 71.

125. See Kirsten Grind, Sam Schechner, Robert McMillan, and John West, "How Google Interferes with Its Search Algorithms and Changes Your Results," *Wall Street Journal*, Nov. 15, 2019, https:www.wsj,com/articles/how-google-interferes-with-its-search-algorithms-and-changes-your-results-11573823753.

126. See generally Stephen Davies, "Decoding the Social Media Algorithms: A Guide for Communicators," https://www.stedavies.com/social-mediaalgorithms-guide, last visited Sept. 21, 2018.

127. See Tony Chapelle, "Facebook, Twitter and Social Media in a Risk Vise," Agenda, Jan. 22, 2018, http://agendaweek.com/pc/1858614/317383.

128. U.S. Postal Service, "Postal Rates for Periodicals: A Narrative History," https://about.usps.com/who-we-are/postal-history/periodicals-postage-history.htm. See Chapter 2.

129. "Financing, Import, and Export," Media Today: Japan, South African, United States, Pennsylvania State University, Apr. 28, 2014, https://sites.psu.edu/yokuokimasu/2014/04/28/financing-import-export.

130. Interview with Rasmus Nielsen, *Media Apocalypse*, episode 3, June 11, 2020, YouTube, https://youtu.be/EUon6QjBIGk.

131. Elizabeth Jensen, "PBS Showed the Future. What Does Its Own Look Like?" *New York Times*, Oct. 25, 2020, https://www.nytimes.com/2020/10/13/arts/television/pbs-future.html

132. "Congressman Tim Ryan Urges House Speaker Pelosi to Provide Assistance to Local Media Outlets and Journalists," What They Think, Apr. 20, 2020, http://whattheythink.com/news/100465-congressman-tim-ryan-urges-house-speaker-pelosi-provide-assistance-local-media-outlets-journalists.

133. Victor Pickard, "Journalism's Market Failure Is a Crisis for Democracy," *Harvard Business Review*, Mar. 12, 2020, https://hbr.org/2020/03/journalisms-market-failure-is-a-crisis-for-democracy. ProPublica shows how digital resources can enhance journalism. See Lena V. Groeger et al., "What Parler Saw During the Attack on the Capitol," *ProPublica*, Jan. 17, 2021, https://projects.propublica.org/parler-capitol-videos/ (journalists scraped data and videos from social media site Parler before it went dark and produced story and information useable by law enforcement about the Jan. 6, 2021 assault on the U.S. Capitol).

134. Catherine Buni, "4 Ways to Fund—and Save—Local Journalism," Nieman Reports, May 7, 2020, https://niemanreports.org/articles/4-ways-to-fund-and-save-journalism; Leonard Downie Jr. and Michael Schudson, "The Reconstruction of American Journalism (2009)," Media Impact Report (2019).

135. One example is the Medill School of Journalism's Local News Initiative, supported by individuals and by a Google Innovation Challenge Grant. Erin Karter, "Bolstering Local News," Northwestern University, Fall 2020.

136. Ken Doctor, "Newsonomics: The New Knight-Lenfest Initiative Gives a Kick in the Pants to America's Metro Newspapers," Nieman Lab, Feb. 13, 2017, http://www.niemanlab.org/2017/02/newsonomics-the-new-knight-lenfest-initiative-gives-a-kick-in-the-pants-to-americas-metro-newspapers; Aude White, "New York Magazine Partners with The City, a Nonprofit Digital News Start-Up," *New York Magazine*, Sept. 26, 2018, http://nymag.com/press/2018/09/new-york-magazine-partners-with-news-start-up-the-city.html; Tony Proscio, "Out of Print: The Case for Philanthropic Support for Local Journalism in a Time of Market Upheaval," Revson Foundation, Jan. 31, 2018, http://revsonfoundation.org/download/publications/Out-of-Print-Report-Tony-Proscio.pdf.

137. Magda Konieczna, *Journalism Without Profit: Making News When the Market Fails* (New York: Oxford University Press, 2018).

138. Paul Romer, "A Tax That Could Fix Big Tech," *New York Times*, May 6, 2019, https://www.nytimes.com/2019/05/06/opinion/tax-facebook-google.html.

139. Ethan Zuckerman, "The Case for Digital Public Infrastructure," Knight First Amendment Institute, Jan. 17, 2020, https://knightcolumbia.org/content/the-case-for-digital-public-infrastructure; Emily Bell, "How

Mark Zuckerberg Could Really Fix Journalism," *Columbia Journalism Review*, Feb. 21, 2017, https://wwwcjr.org/tow_center/mark-zuckerberg-facebook-fix-journalism.php.

140. Buni, "4 Ways to Fund—and Save—Local Journalism." Taxing a specific industry because of its market power or other qualities is lawful if the tax is not related to suppression of expression. *Minneapolis Star & Tribune Co. v. Minn. Comm'r Revenue*, 460 U.S. 575, 585 (1983).

141. Steve Waldman, "A Government Fund to Help Journalism . . . That Wouldn't Corrupt Journalism," Poynter, May 20, 2020, https://www.poynter.org/business-work/2020/a-government-fund-to-help-journalism-that-wouldnt-corrupt-journalism.

142. Matthew C. Nisbet, John P. Wihbey, Silje Kristiansen, and Aleszu Bajak, "Funding the News: U.S. Foundations and Nonprofit Media," Shorenstein Center, Harvard University, and Northeastern University, June 2018, https://camd.northeastern.edu/fundingthenews/#_ga=2.216436246.1862563578.1587003732-868626010.1587003732.

143. See Public Media Alliance, "What Is Public Service Media?," https://www.publicmediaalliance.org/about-us/what-is-psm; European Public Broadcast Union, "Public Service Values, Editorial Principles and Guidelines," https://www.ebu.ch/publications/public-service-values-editrial.

144. See generally Mike Gonzalez, "Is There Any Justification for Continuing to Ask Taxpayers to Fund NPR and PBS?," Knight Foundation, 2017, https://www.knightfoundation.org/public-media-white-paper-2017-gonzalez. For more on local news issues, see "Local Journalism in the Pacific Northwest: Why It Matters, How It's Evolving, and Who Pays for It," University of Oregon School of Journalism and Communications, 2017, https://papers.ssrn.com/sol3/papers.cfm?abstract_id=3045516; see also Molly de Aguiar and Josh Stearns, "Lessons Learned from the Local News Lab," Medium, Feb. 2016, https://medium.com/the-local-news-lab/tagged/lessonslearned.

145. Buni, "4 Ways to Fund—and Save—Local Journalism." A majority of Republicans surveyed and an even larger percentage of Democrats supported funding for public broadcasting in 2017.

146. Generally Blair Levin, *Public Media at 50: What's Next for the Information Commons?*, Knight Foundation, 2017, https://www.knightfoundation.org/public-mediawhite-paper-2017-levin.

147. Adam Ragusea, "Topple the Towers: Why Public Radio and Television Stations Should Radically Reorient Toward Digital-First Local News, and How They Could Do It," Knight Foundation, 2017, https://www.knightfoundation.org/public-media-white-paper-2017-ragusea.

148. See Mark Jacob, "Illinois Would Create Local Journalism Task Force," Northwestern University, https://localnewsinitiative.northwestern.edu/posts/2020/03/13/illinois-task-force/index.html.

149. See James T. Hamilton, "Public Affairs: What the Invisible Hand of the News Market Leaves All Too Invisible," Current, May 17, 2010, https://current.org/2010/05/public-affairs-what-the-invisible-hand-of-the-news-market-leaves-all-too-invisible.

150. See Asa Briggs, *The History of Broadcasting in the United Kingdom* (London: Oxford University Press, 1995); Charlotte Higgins, *This New Noise: The Extraordinary Birth and Troubled Life of the BBC* (London: Faber and Faber, 2015); Melvyn Bragg, "*This New Noise* Review—An Excellent and Insightful History of the BBC," *The Guardian*, June 15, 2015, https://www.theguardian.com/books/2015/jun/15/this-new-noise-review-charlotte-higgins-bbc-melvyn-bragg-extraordinary-birth-and-troubled-life.

151. His speech, invoking the public interest and criticizing commercial broadcasting for creating a "vast wasteland," produced controversy in the industry but secured broad public support. Jane Hall, "A Look Back at 30 Years of TV's 'Vast Wasteland,'" *Los Angeles Times*, May 9, 1991, https://www.latimes.com/archives/la-xpm-1991-05-09-ca-1990-story.html. See Frontline: "Interviews: Newton Minow," March, 31, 2006, https://www.pbs.org/wgbh/pages/frontline/newswar/interviews/minow.html: "Why do we have public libraries? Why do we have public parks? Why do we have public universities? Why do we have public hospitals? Why do we do that? We do that because there's a market failure. The market does not serve everybody fairly, so we therefore decide that we will have a noncommercial alternative for people."

152. Robert Morrow, *Sesame Street and the Reform of Children's Television* (Baltimore: Johns Hopkins University Press, 2008).

153. Joy Mayer, "Who Trusts—and Pays for—the News? Here's What 8,728 People Told Us," Reynolds Journalism Institute, July 27, 2017, https://www.rjionline.org/stories/who-trusts-and-pays-for-the-news-heres-what-8728-people-told-us. See Mike Hale, "PBS Is Still TV's Best Path to Better Citizenship," *New York Times*, Oct. 13, 2020, https://www.nytimes.com/2020/10/13/arts/television/pbs-american-life.html.

154. Mike Rispoli, "Why the Civic Info Consortium Is Such a Huge Deal," Free Press, Jan. 16, 2020, https://www.freepress.net/our-response/expert-analysis/insights-opinions/why-civic-info-consortium-such-huge-deal.

155. Nic Newman with Richard Fletcher, Anne Schulz, Simge Andı, and Rasmus Nielsen, *Reuters Institute Digital News Report 2020*, Reuters Institute, June 2020, 9–10, https://reutersinstitute.politics.ox.ac.uk/sites/default/files/2020-06/DNR_2020_FINAL.pdf.

156. Ganesh Sitaraman and Anne L. Alstott, *The Public Option: How to Expand Freedom, Increase Opportunity, and Promote Equality* (Cambridge, MA: Harvard University Press, 2019), 224: "As Senator Angus King (an independent from Maine) has said, 'Failure to provide broadband to rural areas of America is a death sentence for those communities.'"

157. Brad Plumer, "Why Exactly Should the Government Fund PBS and NPR?," *Washington Post*, Oct. 10, 2012, https://www.washingtonpost.com/news/wonk/wp/2012/10/10/why-exactly-should-the-government-fund-pbs-and-npr/?noredirect=on&utm_term=.27c555347878. Arguments for and against government ownership of airwaves to serve the public interest and raise quality reflect competing views about what is more threatening, big business or big government. See Laura Weinrib, *The Taming of Free Speech: America's Civil Liberties Compromise* (Cambridge, MA: Harvard University Press, 2016), 311–328.

158. See "PBS and WGBH to Provide At-Home Learning Programs for Students and Educators Nationwide During School Closures Through WORLD Channel," WGBH, Mar. 26, 2020, https://www.wgbh.org/foundation/press/pbs-and-wgbh-to-provide-at-home-learning-programs-for-students-and-educators-nationwide-during-school-closures-through-world-channel. See also Newton Minow, "A Vaster Wasteland," *The Atlantic*, Apr. 2011, https://www.theatlantic.com/magazine/archive/2011/04/a-vaster-wasteland/308418.

159. William F. Fore, "In Defense of Public Broadcasting," *Christian Century*, July 5–12, 1995, 668–679, https://www.religion-online.org/article/in-defense-of-public-broadcasting.

160. Sue Gardner, "Public Broadcasting: Its Past and Its Future," Knight Foundation, Dec. 2007, https://kf-site-production.s3.amazonaws.com/media_elements/files/000/000/115/original/Topos_KF_White-Paper_Sue-Gardner_V2.pdf.

161. Diane Coyle, "We Need a Publicly Funded Rival to Facebook and Google," *Financial Times*, July 9, 2018, https://www.ft.com/content/d56744a0-835c-11e8-9199c2a4754b5a0e; Zoe Schiffer, "'Filter Bubble' Author Eli Pariser on Why We Need Publicly Owned Social Networks," The Verge, Nov 12, 2019, https://www.theverge.com/interface/2019/11/12/20959479/eli-pariser-civic-signals-filter-bubble-q-a.

162. See Kyle Langvardt, "A New Deal for the Online Public Sphere," *George Mason Law Review* 26 (2018): 52.

163. See Diane Coyle, "Diane Coyle Outlines Her Vision for a 21st Century BBC," *The Guardian*, June 24, 2014, https://www.theguardian.com/media/2014/jun/24/diane-coyle-lecture-vision-21st-century-bbc-full-text; Mike Gonzales, "Is There Any Justification for Continuing to Ask Taxpayers to Fund NPR and PBS?," Knight Foundation, https://www.knightfoundation.org/public-media-white-paper-2017gonzalez, last visited Oct. 25, 2018.

164. See "Losing the News: The Decimation of Local Journalism and the Search for Solutions," PEN America, Nov. 20, 2019, https://pen.org/wp-content/uploads/2019/12/Losing-the-News-The-Decimation-of-Local-Journalism-and-the-Search-for-Solutions-Report.pdf.

165. Waldman, "A Government Fund to Help Journalism."

166. McChesney and Nichols, *The Death and Life of American Journalism*, 84–85.

167. See Emily Bonilla, "Why Media Literacy Education Matters in the Era of Fake News," TNTP, Dec. 13, 2016, https://tntp.org/blog/post/why-media-literacy-education-matters-in-the-era-of-fake-news; "What Is Digital Literacy?," Common Sense Media, https://www.commonsensemedia.org/news-and-medialiteracy/what-is-digital-literacy. Children often feel neglected, misrepresented, or depressed by news they encounter in varied media, and also have trouble distinguishing fake news stories from real ones. Michael B. Robb, "News and America's Kids: How Young People Perceive and Are Impacted by the News," Common Sense Media, Mar. 2017, https://www.commonsensemedia.org/research/news-and-americas-kids.

168. See generally Joseph Kahne, Nam-Jin Lee, and Jessica Timpany Feezell, "Digital Media Literacy Education and Online Civic and Political Participation," *International Journal of Communications* 6 (2012): 1.

169. See Center for Media Literacy, "10 Benefits of Media Literacy," https://www.medialit.org/reading-room/10-benefits-media-literacy-education; "News and Media Literacy Resource Center," Common Sense Education, https://www.commonsense.org/education/news-media-literacy-resource-center; "Core Principles," National Association for Media Literacy Education, https://namle.net/publications/core-principles. See "Facing History and Ourselves," https://www.facinghistory.org/educator-resources/current-events/plan-ahead-current-events-teacher-checklist.

170. Marshall Allen, "I'm an Investigative Journalist. These Are the Questions I Asked About the 'Viral Plandemic' Video," ProPublica, May 9, 2020, https://www.propublica.org/article/im-an-investigative-journalist-these-are-the-questions-i-asked-about-the-viral-plandemic-video.

171. See generally Katie Benner, "Snapchat Remakes Itself, Splitting the Social from the Media," *New York Times*, Nov. 29, 2017, https://www.nytimes.com/2017/11/29/technology/snapchat-redesign-social-media.html. Cultivating interest in and attention to high-quality news is a serious problem made worse by the addictive nature of social media. See Ethan Zuckerman, "Four Problems for News and Democracy," Medium, Sept. 24, 2018, https://medium.com/trust-media-and-democracy/we-know-the-news-is-incrsrisi-5d1c4fbf7691.

CHAPTER FIVE

1. Victor Pickard, *Democracy Without Journalism? Confronting the Misinformation Society* (New York: Oxford University Press, 2019); Michael Luo, "The Fate of the News in the Age of the Coronavirus," *New Yorker*, March 29, 2020, https://www.newyorker.com/news/annals-of-communications/the-fate-of-the-news-in-the-age-of-the-coronavirus.

2. See generally Dara Purvis, "Alexander Meiklejohn," in *Encyclopedia of the First Amendment*, ed. John R. Vile et al. (Washington, DC: CQ Press, 2009), 734.

3. Alexander Meiklejohn, *Free Speech and Its Relation to Self-Government* (New York: Harper, 1948), 85.

Index

. . .

For the benefit of digital users, indexed terms that span two pages (e.g., 52–53) may, on occasion, appear on only one of those pages.

INDEX